SC

SCOPE 26: Meth... Non-Human Biota, 1985, 712 pp

SCOPE 27: Climate Impact Assessment: Studies ... Climate and Society, 1985, 650 pp

SCOPE 28: Environmental Consequences of Nuclear War
Volume I Physical and Atmospheric Effects, 1986, 400 pp
Volume II Ecological and Agricultural Effects, 1985, 563 pp

SCOPE 29: The Greenhouse Effect, Climatic Change, and Ecosystems, 1986, 574 pp

SCOPE 30: Methods for Assessing the Effects of Mixtures of Chemicals, 1987, 928 pp

SCOPE 31: Lead, Mercury, Cadmium and Arsenic in the Environment, 1987, 384 pp

SCOPE 32: Land Transformation in Agriculture, 1987, 552 pp

SCOPE 33: Nitrogen Cycling in Coastal Marine Environments, 1988, 478 pp

SCOPE 34: Practitioner's Handbook on the Modelling of Dynamic Change in Ecosystems, 1988, 196 pp

SCOPE 35: Scales and Global Change: Spatial and Temporal Variability in Biospheric and Geospheric Processes, 1988, 376 pp

SCOPE 36: Acidification in Tropical Countries, 1988, 424 pp

SCOPE 37: Biological Invasions: A Global Perspective, 1989, 528 pp

SCOPE 38: Ecotoxicology and Climate with Special Reference to Hot and Cold Climates, 1989, 432 pp

SCOPE 39: Evolution of the Global Biogeochemical Sulphur Cycle, 1989, 224 pp

SCOPE 40: Methods for Assessing and Reducing Injury from Chemical Accidents, 1989, 320 pp

SCOPE 41: Short-Term Toxicity Tests for Non-Genotoxic Effects, 1990, 386 pp

SCOPE 42: Biogeochemistry of Major World Rivers, 1990

SCOPE 43: Stable Isotopes in the Assessment of Natural and Anthropogenic Sulphur in the Environment, 1990

SCOPE 44: Introduction of Genetically Modified Organisms into the Environment, 1990, 224 pp

Funds to meet SCOPE expenses are provided by contributions from SCOPE Committees, an annual subvention from ICSU (and through ICSU, from UNESCO), an annual subvention from the French Ministère de l'Environment, contracts with UN Bodies, particularly UNEP, and grants from Foundations and industrial enterprises.

SCOPE 44

Introduction of Genetically Modified Organisms into the Environment

SCOPE 44

Introduction of Genetically Modified Organisms into the Environment

Edited by
HAROLD A. MOONEY
Department of Biological Sciences, Stanford University, Stanford, California, USA

GIORGIO BERNARDI
Laboratoire de Génétique Moléculaire, Institut Jacques Monod, Paris, France

Published on behalf of the Scientific Committee on Problems of the Environment (SCOPE) and the Scientific Committee on Genetic Experimentation (COGENE) of the International Council of Scientific Unions (ICSU)

by

JOHN WILEY & SONS
Chichester · New York · Brisbane · Toronto · Singapore

Other Wiley Editorial Offices

John Wiley & Sons, Inc., 605 Third Avenue
New York, NY 10158-0012, USA

Jacaranda Wiley Ltd, GPO Box 859, Brisbane,
Queensland 4001, Australia

John Wiley & Sons (Canada) Ltd, 22 Worcester Road,
Rexdale, Ontario M9W 1L1, Canada

John Wiley & Sons (SEA) Pte Ltd, 37 Jalan Pemimpin 05–04,
Block B, Union Industrial Building, Singapore 2057

Library of Congress Cataloging-in-Publication Data:

Introduction of genetically Modified organisms into the environment/edited by Harold A. Mooney and Giorgio Bernardi.
p. cm. — (SCOPE ; 44)
Includes bibliographical references.
ISBN 0 471 92677 9
1. Genetic engineering — Safety measures. 2.Genetic engineering — Environmental aspects. I. Mooney, Harold A. II. Bernardi, Giorgio. III. International Council of Scientific Unions. Scientific Committee on Problems of the Environment. IV. Committee on Genetic Experimentation. V. Series: SCOPE report ; 44.
TP248.6.I6 1990
574.5′223—dc20 90-35076
CIP

British Library Cataloguing in Publication Data:

Introduction of genetically Modified organisms into the environment.
1. Organisms. Genetic engineering. Environmental aspects I. Mooney, H.A. (Harold A.) II. Bernardi, Giorgio III. Series
575.1

ISBN 0 471 92677 9

Typeset by Photo·graphics, Honiton, Devon.
Printed in Great Britain by Biddles Ltd, Guildford, Surrey

International Council of Scientific Unions (ICSU)
Scientific Committee on Problems of the Environment (SCOPE)

SCOPE is one of a number of committees established by the non-governmental group of scientific organizations, the International Council of Scientific Unions (ICSU). The membership of ICSU includes representatives from 67 National Academies of Science, 20 International Unions, and 33 other bodies called Associates. To cover multidisciplinary activities which include the interests of several unions, ICSU has established 10 Scientific Committees, of which SCOPE is one. Currently representatives of 36 member countries and 21 Unions, Scientific Committees and Associates participate in the work of SCOPE, which directs particular attention to the needs of developing countries. SCOPE was established in 1969 in response to the environmental concerns emerging at the time: ICSU recognized that many of these concerns required scientific inputs spanning several disciplines and ICSU Unions. SCOPE's first task was to prepare a report on Global Environmental Monitoring (SCOPE 1, 1971) for the UN Stockholm Conference on the Human Environment.

The mandate of SCOPE is to assemble, review, and assess the information available on man-made environmental changes and the effects of these changes on man: to assess and evaluate the methodologies of measurement of environmental parameters; to provide an intelligence service on current research; and by the recruitment of the best available scientific information and constructive thinking to establish itself as a corpus of informed advice for the benefit of centres of fundamental research and of organizations and agencies operationally engaged in studies of the environment.

SCOPE is governed by a General Assembly, which meets every three years. Between such meetings its activities are directed by the Executive Committee.

R.E. Munn
Editor-in-Chief
SCOPE Publications

Executive Secretary: V. Plocq

Secretariat: 51 Bld de Montmorency
75016 PARIS

Contents

List of Contributors

Werner Arber
Department of Microbiology, Biozentrum, University of Basel, Klingelbergstrasse 70, CH-4056 Basel, Switzerland

Giorgio Bernardi
Laboratoire de Génétique Moléculaire, Institut Jacques Monod, 2, Place Jussieu, 75005 Paris, France

Susana Brom
Departamento de Genética Molecular, Centro de Investigación sobre Fijación de Nitrógeno, Universidad Nacional Autónoma de México, Apartado Postal 565-A, Cuernavaca, Morelos, Mexico

Allan Campbell
Department of Biological Sciences, Stanford University, Stanford, CA 94305, USA

A. Caplan
Institute of Genetics, University of Gent, B-9000 Gent, Belgium

Baya Cherif-Zahar
Laboratoire de Génétique Moléculaire, Institut Jacques Monod, 2, Place Jussieu, 75005 Paris, France

Freddy Bugge Christiansen
Department of Ecology and Genetics, University of Aarhus, DK-8000 Aarhus, Denmark

Michael J. Crawley
Department of Pure and Applied Biology, Imperial College, Silwood Park, Ascot, Berkshire SL5 7PY, UK

Guillermo Dávila
Departamento de Genética Molecular, Centro de Investigación sobre Fijación de Nitrógeno, Universidad Nacional Autónoma de México, Apartado Postal 565-A, Cuernavaca, Morelos, Mexico

James A. Drake
Department of Zoology and the Graduate Program in Ecology, University of Tennessee, Knoxville, TN 37996, USA

Daryl F. Dwyer
Department of Microbiology, GBF-National Research Center for Biotechnology, Braunschweig, Federal Republic of Germany

Robert M. Faust
United States Department of Agriculture, Agriculture Research Service, Beltsville, MD 20705, USA

Frank Fenner
The John Curtin School of Medical Research, The Australian National University, Canberra 2601, Australia

Margarita Flores
Departamento de Genética Molecular, Centro de Investigación sobre Fijación de Nitrógeno, Universidad Nacional Autónoma de México, Apartado Postal 565-A, Cuernavaca, Morelos, Mexico

Regina Goursot
Laboratoire de Génétique Moléculaire, Institut Jacques Monod, 2, Place Jussieu, 75005 Paris, France

René Goursot
Laboratoire de Génétique Moléculaire, Institut Jacques Monod, 2, Place Jussieu, 75005 Paris, France

Kunthala Jayaraman
Center for Biotechnology, Anna University, Madras 600 025, India

David T. Kingsbury
Department of Microbiology, The George Washington Medical Center, Washington, DC 20037, USA

Simon A. Levin
Charles H. Alexander Professor of Biological Sciences, Section of Ecology and Systematics, and Director, Center for Environmental Research, 347 Corson Hall, Cornell University, Ithaca, NY 14853, USA

Esperanza Martinez
Departamento de Genética Molecular, Centro de Investigación sobre Fijación de Nitrógeno, Universidad Nacional Autónoma de México, Apartado Postal 565-A, Cuernavaca, Morelos, Mexico

Roberta Melis
Laboratoire de Génétique Moléculaire, Institut Jacques Monod, 2, Place Jussieu, 75005 Paris, France

Harold A. Mooney
Department of Biological Sciences, Stanford University, Stanford, CA 94305, USA

Rafael Palacios
Departamento de Genética Molecular, Centro de Investigación sobre Fijación de Nitrógeno, Universidad Nacional Autónoma de México, Apartado Postal 565-A, Cuernavaca, Morelos, Mexico

Daniel Piñero
Departamento de Ecología, Instituto de Biologia, Universidad Nacional Autónoma de México, Apartado Postal 41-746, Cuernavaca, Morelos, Mexico

Edda Rayko
Laboratoire de Génétique Moléculaire, Institut Jacques Monod, 2, Place Jussieu, 75005 Paris, France

David Romero
Departamento de Genética Molecular, Centro de Investigación sobre Fijación de Nitrógeno, Universidad Nacional Autónoma de México, Apartado Postal 565-A, Cuernavaca, Morelos, Mexico

Anna Marie Skalka
Fox Chase Cancer Center, Institute for Cancer Research, 7701 Burholme Avenue, Philadelphia, PA 19111, USA

Kenneth N. Timmis
Department of Microbiology, GBF-National Research Center for Biotechnology, Braunschweig, Federal Republic of Germany

M. Van Montagu
Institute of Genetics, University of Gent, B-9000 Gent, Belgium

Discussion Participants

R.M. Anderson
Department of Pure and Applied Biology, Imperial College, Prince Consort Road, London SW7 2BB, UK

U. Bertazzoni
CEC, Rue de la Loi 200, 1049 Brussels, Belgium

S.N. Cohen
Department of Genetics, Stanford University, Stanford, CA 94305, USA

R.R. Colwell
Maryland Biotechnology Institute, University of Maryland, College Park, MD 20742, USA

P.J. Regal
Department of Ecology and Behavioral Biology, University of Minnesota, Minneapolis, MN 55455, USA

V. Sgaramella
Institute of Genetics, University of Pavia, Via Abbiategrasso 207, Pavia 27100, Italy

Preface

The Scientific Committee on Problems of the Environment (SCOPE) and the Committee on Genetic Experimentation(COGENE) co-sponsored a meeting of specialists in molecular biology, population genetics, and ecology to consider the issues related to the introduction of genetically modified organisms into the environment. The charge to this group was to provide the necessary scientific background in order to make a statement on the potential benefits and hazards of introductions of bio-engineered organisms. This volume contains first the statement and then the background from which it is derived.

The participants of the meeting included those who have been central in the revolution of thought and techniques that have characterized the development of molecular biology as well as ecologists who have actively participated in an analysis of the fate and environmental consequences of introduced organisms.

At the time of the meeting there had been considerable discussion at national levels concerning the consequences of the 'release' of genetically engineered organisms, but there had been less international discourse. Further, the attempts at dialogue between ecologists and molecular biologists had not been highly successful, in part because of meeting formats. This meeting, held at the Rockefeller Conference Center in Bellagio, Italy, was designed to have international participation and to utilize a theme that was familiar and central to all of the disciplines involved. This core theme was the evolutionary and environmental dimensions of the release of bio-engineered organisms.

Support of the A.W. Mellon and Rockefeller Foundations is gratefully acknowledged.

H.A. Mooney
G. Bernardi

Joint SCOPE/COGENE Statement

INTRODUCTION

Genetic change is a fundamental biological phenomenon and is the basic process of evolution. It has been exploited by humans for their own purposes throughout history. Recent methodological developments, such as advances in the application of recombinant DNA (rDNA) techniques, have expanded these capabilities and provide the opportunity for addressing a wide variety of pressing human needs including solutions to environmental problems. Efforts to reduce pollution, to combat microbial infection, and to improve food production may require applications that involve the large-scale introduction of genetically modified organisms into the environment.

Any human intervention into the environment carries an element of risk. The natural response to this uncertainty is to proceed cautiously, but one must weigh any risks that might attend the implementation of the methodology against the benefits and against any risks that might be associated with delaying implementation.

The environmental introduction of any organisms, modified or unmodified, should be undertaken within a framework that maintains appropriate safeguards for the protection of the environment and human health while not discouraging innovation. This implies that an ecological perspective is essential to the assessment of risks associated with any introduction.

There is considerable experience in the introduction of organisms into the environment. Familiar examples include vaccines, agricultural varieties, and biological control. The positive and negative lessons learned from these experiences provide a basis for the rational development of safeguards. Nevertheless, these is much that still needs to be known about such topics as systematics; community structure; the factors governing the survival, growth, and spread of populations of introduced organisms; and techniques for monitoring.

The development of new techniques also opens up new and exciting vistas for basic research, many of them involving close collaboration between molecular biologists, ecologists, and evolutionists. The new techniques will help us to understand the genetic structure of populations and to develop improved methods in applied ecology; ecological and epidemiological perspectives may serve to elucidate fundamental theoretical aspects of molecular biology, such as the population biology of plasmids and transposable elements.

RISKS

In view of the great potential of new technologies for addressing environmental and other problems, and because most introductions of modified organisms are likely to represent low or negligible ecological risk, generic arguments against the use of new genetic methodologies must be rejected. Indeed, the spectrum of available tools represents an evolving and expanding continuum, which includes conventional methods, rDNA techniques, and others. While much attention has been focused on the methods used to modify organisms, it is the products of these technologies and the uses to which they will be put that should be the objects of attention, rather than the particular techniques employed to achieve those ends.

Similarly, one must reject generic safety arguments based on the assertion that all introductions must have occurred some time during the course of evolution. Therefore, each introduction of an organism, whether modified or not, must be judged on its own merits, within the context of the scale of the application and the possible environmental costs and benefits.

Size, geographical scale, and frequency of introduction are among the factors that are important in determining whether a particular introduction will become established or spread. Therefore, small-scale field testing involves different considerations than does large-scale (e.g. commercial) application. This is not to suggest that small-scale testing should be exempt from examination for regulation, but simply that any risks are likely to be much smaller and more easily managed than those for large-scale applications. Nor does this suggest that all large-scale applications will be problematical, since we have many examples to the contrary, including vaccines, biological control methods, and the use of rhizobia in agriculture.

One must bear in mind that the greatest impact on the biosphere is through human activity. For example, deforestation, the widespread use of antibiotics, vaccines, herbicides, and pesticides, or the widespread introduction of a single type of organism in agriculture has led to a loss of genetic diversity. Such loss has had dramatic and unfortunate consequences for human survival because it has led to environmental degradation, loss of stability, and a depletion of biological resources that are valuable for food, fibre, medicines, and other purposes.

As the ability to manipulate our environment increases, so must an awareness of these problems and of the need to exercise our power wisely.

RISK ASSESSMENT

As already discussed, the risks of making a specific introduction must be weighed against the perceived benefits and the risks of not making the introduction. This balancing is part of risk management. This section addresses only the assessment of risks of introduction, but it is important that it be integrated into the management framework.

The properties of the introduced organism and its target environment are the key features in the assessment of risk. Such factors as the demographic characterization of the introduced organisms; genetic stability, including the potential for horizontal transfer or outcrossing with weedy species; and the fit of the species to the physical and biological environment. The scale and frequency of the introductions are important related factors. These considerations apply equally to both modified or unmodified organisms, and, in the case of modified organisms, they apply independently of the techniques used to achieve modification; that is it is the organism itself, and not how it was constructed, that is important.

Each proposed introduction must be treated on its own merits, but this does not suggest that each needs to be considered *de novo*. As experience accumulates with particular kinds of introductions in particular environments, more generic approaches to these classes of introductions can be developed. The bases for classification should be refined continually, providing a set of criteria that will allow any proposed introductions judged to be innocuous to be carried out speedily and those judged to be problematical to be given the attention they deserve.

It is important to note that generalizations developed for particular groups of organisms (e.g. microorganisms) cannot be extended automatically to other groups (e.g. plants), which may have very different genetic and demographic characteristics, dispersal and reproductive mechanisms, and trophic positions.

In the development of assessment procedures, the potential for containment, monitoring, and mitigation must receive consideration. In this regard, it should generally be assumed that the likelihood of being able to recall introduced organisms that have escaped containment is very small.

CODA

The new genetic methodologies add additional tools to the spectrum of techniques available to improve the human condition. These will lead to rapid improvements in the development of ecologically sound approaches to agriculture and to environmental management, and in the acquisition of a better understanding of biological systems. These benefits in their various forms should be available to everyone, and know no national boundaries. It is therefore essential that the broadest possible international cooperation and datasharing be supported by all countries. Each improvement in our capacity to modify the environment carries with it responsibilities to use that capability wisely, with special attention given to the importance of maintaining biological diversity.

Section 1

HISTORY OF CONCERNS REGARDING GENETIC MANIPULATION

Introduction of Genetically Modified Organisms into the Environment. Edited by Harold A. Mooney and Giorgio Bernardi

1 Risk Assessment for Genetic Experimentation and Application

ANNA MARIE SKALKA
Fox Chase Cancer Center, Institute for Cancer Research, Philadelphia, Pennsylvania, USA

1.1 LABORATORY RESEARCH AND THE COMMITTEE FOR GENETIC EXPERIMENTATION (COGENE)

1.1.1 THE MANDATE

The Working Group on Risk Assessment was one of the first subcommittees to be formed by COGENE to evaluate the impact of recombinant DNA technology. This group was very active during the late 1970s at a time of great public anxiety concerning the use of this technology in the laboratory. The intensity of anxiety varied a great deal in different countries, as did the public response. Since the presumed ramifications of such work transcended geographical boundaries, it seemed appropriate to provide international scientific evaluations to the public discussions.

The mandate for this Working Group, as for many of the Scientific Committees of the International Council of Scientific Unions (ICSU), was to gather information from countries throughout the world. A second objective was to identify areas from which additional useful information might be obtained and, if necessary, support efforts to produce or collate these data. As a starting point, the Working Group collected information, in the form of a questionnaire, from 21 individual nations or organizations throughout the world. In addition, committee members attended many national and international meetings, examined relevant publications, and reviewed various public reports.

1.1.2 COGENE REPORTS

In 1978 the Working Group produced its first report to COGENE which was organized in two volumes. The second of these contained documentation in support of the summary statements provided in a smaller first volume. In its summary, the Working Group identified the three most commonly articulated concerns related to the use of recombinant DNA technology.

Introduction of Genetically Modified Organisms into the Environment. Edited by Harold A. Mooney and Giorgio Bernardi

1. Genetically modified microorganisms might spread from the laboratory into the environment and disrupt existing ecological equilibria.
2. Genetically modified organisms might produce some toxic or noxious substance or otherwise cause disease.
3. By using this new technology, scientists may be crossing some hypothetical barrier to DNA exchange between eukaryotes and prokaryotes and thus affect the pathogenicity or dispersion of pathological agents.

It is important to emphasize that in the late 1970s these concerns were focused primarily on the use of common laboratory strains of bacteria, such as *Escherichia coli* or *Bacillus subtilis*, as host organism for recombinant plasmids. It was primarily towards such applications that our risk assessment was directed.

As stated in the Summary report (Skalka *et al.*, 1978), after consideration of all the available evidence, the Working Group concluded that there were no scientific findings to justify the concerns identified above. No risk unique to recombinant DNA research was identified; recombinations of the type made possible by these techniques could occur in nature. Furthermore, evaluation of the organism to be used in the vast majority of the research (i.e. *E. coli* strain K12) showed that it was essentially harmless and that insertion of foreign DNA into its genome could not alter this property. We speculated further that '...with few exceptions it seems likely that the same will prove true of other (harmless) host bacteria.'

The Working Group did, however, identify an area in which additional information would be useful, namely the transmissibility of *E. coli* and its plasmids. Towards this end we were able to solicit the assistance of three well-known scientists who had been testing the effect of feeding *E. coli* K12 to human volunteers and were also involved in routine monitoring of their laboratory workers. These investigators, Dr Stanley Falkow of the United States and Drs M. Richmond and E.S. Anderson of the United Kingdom, agreed to collate the results of their work in a second report for COGENE (Skalka *et al.*, 1978, and Skalka, 1979, App. II). These results showed that there was no acquisition of bacteria or transmissible plasmids either in individuals who ingested *E. coli* K12 or in laboratory workers who had been working with plasmid containing *E. coli* K12 strains over the years. These findings were of special importance because they reflected what happened in real working situations where genetically modified organisms were handled with no special precautions. They served to restore confidence in the scientists' ability to predict the outcome of a situation based on available knowledge of the organism involved. It seemed reasonable, therefore, that similar facts relating to other potential systems should prove equally predictive and that as experience progressed fewer risk assessments should be required.

1.1.3 THE IMPACT OF COGENE'S EFFORTS

The reports of the Working Group were distributed to various national and international organizations. They served as a resource for executive action in several national bodies. For example, the reports and material cited therein were used as part of the justification for the eventual relaxation of the 'NIH Guidelines' for work with recombinant DNA. The reports also had a significant impact in the United Kingdom by helping to curb efforts to impose the excessively restrictive legislation being considered at the time.

1.2 A CURRENT ROLE FOR COGENE WITH THE SCIENTIFIC COMMITTEE ON PROBLEMS OF THE ENVIRONMENT (SCOPE)

1.2.1 PRACTICAL APPLICATIONS AND THEIR ENVIRONMENTAL IMPACT

At the time of the second report to COGENE, it was already clear that the future use of recombinant DNA technology would progress from laboratory experimentation to commercial application. In this summary, the Working Group cautioned that it did

> . . . not consider it justified to extrapolate from experience with *E. coli* to all possible experiments in the future. As technology related to genetic experimentation evolves, there are sure to be operations and applications that we cannot even envision at this time. Thus, it is important that a Committee of scientists, international in scope, continues to follow developments which relate to genetic experimentations, that it continues to collect information to be made broadly available and that it stands ready to sponsor specific projects which may be necesary for further evaluation.

1.2.2 A GLOBAL PERSPECTIVE

Because the global environment is its prime concern, SCOPE'S interests have always transcended national boundaries. Its role in providing information concerning man's impact on the environment and recommendations for appropriate action is well established, highly regarded, and of significant import.

A pressing need in the current context is for scientists, as well as national and international governing bodies, to obtain a realistic appraisal of the impact of the perturbations which might be introduced with the application of genetically modified organisms as compared to the perturbations already experienced. Our biosphere has been and continues to be changed in numerous ways including:

1. The wide-scale use of chemical pesticides.

2. Accumulation of nonbiodegradable products.
3. Toxic wastes that pollute the soil, air, and water.
4. Current agricultural practices which lead to uniformity in the genetic composition of farm animals and crops and loss of biological diversity.
5. Natural biological calamities, such as the current AIDS epidemic.

It seems appropriate to ask how the hypothetical risk associated with currently proposed uses of genetically modified organisms equate with the known risks associated with current practices. This is essential because it is, in large part, public dissatisfaction with the outcome of the latter that has prompted concerns about modern genetics. Industrialized people, healthy and well fed as a result of technological advances, have begun to question whether they have entered into a Faustian bargain with 'science' at the same time as populations in the developing countries are denuding their rain forests in the quest for an adequate food supply. It is important to emphasize that there is no more *intrinsic* danger in the application of modern genetic methods than in the use of modern farm machinery. A perspective on the relative impact of current and proposed pertubations could represent an important contribution to current deliberations.

1.2.3 BIOLOGICAL ALTERNATIVES

One of the most compelling arguments for application of modern genetic technology comes from the promise of environmentally sounder solutions to many of mankind's current needs—for more abundant and available food supplies, for better methods to control and combat disease, and to preserve the environment so that future generations can enjoy an acceptable, even improved, quality of life.

Evaluations that involve both molecular geneticists and environmentalists could help to identify likely applications of genetic technology and establish the most effective and appropriate methods of using them. Considerable experience exists in the introduction of organisms into the environment and this experience can provide some meaningful guideposts. It is also apparent that new genetic tools will open up new opportunities to obtain additional useful knowledge concerning the relationship among various species in the environment. Such knowledge can provide a rational base to the establishment of prudent procedures for future applications.

1.2.4 THE RISK OF SAYING NO

As the world population explodes there is an ever-increasing pressure on the earth's limited resources. Thus, it is appropriate that any intervention into the environment be scrutinized carefully. It is also important in such deliberations to consider the alternatives and to measure the 'hidden' cost of inappropriately restrictive regulations, the cost of saying 'No'. For example:

1. Blocking the progress of simple, carefully controlled risk assessment experiments by various groups with separate political agendas. In such instances, COGENE and SCOPE can play an important role in providing a balanced perspective to policy-making bodies. Surely governmental agencies must have an opportunity to obtain information which will be critical for the establishment of informed policy.
2. The cost, in natural resources, exacted by continued practices such as the massive use of pesticides and ill-conceived agricultural practices. Some of the attendant problems may be alleviated by development of genetically modified resistant plants and higher food-yielding species.
3. The cost in lives if modern genetic tools cannot be used in development and applications of new vaccines for world-wide scourges such as malaria and AIDS. Indeed, were it not for the use of recombinant DNA technology, we would still be unaware of the viral etiology of AIDS and could not have developed the tools for its diagnosis. Many lives have been saved in the last few years alone simply by screening of the blood supply! Could we have afforded to say 'No' in this instance?
4. To most geneticists, and to environmentalists as well, one of the most alarming consequences of modern man's activity has been the destruction of numerous biological species, either directly or indirectly, by the imposition of 'unnatural' selective pressures. This loss in our total genetic heritage means that some valuable traits, and attendant opportunities for solution to the world's problems in the future, may be gone forever. Modern genetic technology, by offering alternatives to selective in-breeding of farm animals, improvement in the food value of crop plants, and new methods for handling toxic wastes, could diminish some of the selective pressure and thus provide more environmentally sound means for satisfying man's immediate needs.

1.3 SUMMARY

It seems clear that our community of scientists has an important responsibility in the continuing debate concerning the introduction of genetically modified organisms into the environment. There is an obvious need for gathering more knowledge, for the objective consideration of information that already exists, and for the formulation of both short-term and long-term solutions to problems of the biosphere. There is an immediate need to provide guidance on how to proceed prudently at the present time without stifling initiative and progress. There is also a need for a broader consideration—one that has as its goal the preservation of our ecological and spiritual heritage for future generations. Together COGENE and SCOPE encompass the multidisciplinary expertise to address these problems. By doing so we can fulfil the principal objective of ICSU which is to encourage international scientific activity for the benefit of all.

REFERENCES

Skalka, A.M. (1979) Second report of the COGENE Working Group on Risk Assessment. In: Morgan, J. and Whelan, W.J. (Eds.) *Recombinant DNA and Genetic Experimentation*, Pergamon Press, Oxford and New York, pp. 211–20.

Skalka, A.M., Bernardi, G. and Sgaramella, V. (1978) First report to COGENE from the Working Group on Risk Assessment. Available from COGENE c/o ICSU Secretariat, 51 Bd. de Montmorency, 75016 Paris, France.

2 Recombinant DNA: Past Lessons and Current Concerns

ALLAN CAMPBELL
Department of Biological Sciences, Stanford University, Stanford, California, USA

I discuss here some general and historical aspects of recombinant DNA concerns and my own involvement with them. This may seem inappropriate to this volume, either because it rehashes ostensibly dead issues or because it includes political and regulatory, as well as scientific, considerations. However, I believe the history is important to understanding our present outlook. Labelling an issue 'dead' implies that it once was alive, whereas much of the recombinant DNA debate was based on notions that were never scientifically viable. The role of competition in limiting the spread of engineered organisms, for example, was scarcely mentioned at the 1975 Asilomar Conference on Recombinant DNA, although most of us would agree that it is central, based on concepts that were in place long before 1975. I have also seen enough discussions of the underlying science to realize how often they can be influenced by some political second guessing of their probable impact; so I would rather put that aspect on the table rather than under it.

2.1 RECOMBINANT DNA CONCERNS: HISTORICAL BACKGROUND

Each of our perspectives on this subject is inevitably conditioned by personal experiences. Briefly, my own participation has been as follows: I attended the Asilomar Conference on Recombinant DNA and chaired the first Stanford University institutional Recombinant DNA Committee the following year. I was a member of the NIH Recombinant Advisory Committee (RAC) from 1977 to 1981 and participated in the COGENE Meeting on Recombinant DNA and Genetic Experimentation held at Wye in 1979. Since 1981 my involvement has been slight, including service on one advisory committee to the Environmental Protection Agency and current membership on the Biosafety Committee of a local genetic engineering company.

Introduction of Genetically Modified Organisms into the Environment. Edited by Harold A. Mooney and Giorgio Bernardi

When I joined RAC, the NIH Guidelines were already in place. As time progressed, I became increasingly conscious that the Guidelines were arbitrary, based more on public relations than on serious science, responsive primarily to fear rather than to danger. The authors of the Guidelines themselves seemed unable to remember or to agree on why or how they had arrived at a particular classification level for a given experiment. Most of my energy was spent working for relaxations of the Guidelines, trying to narrow their focus to any areas where perceptions of danger might have a rational basis. One of my last acts as an RAC member was to sponsor, along with David Baltimore, a proposal that would have removed all but a small fraction of recombinant DNA experiments from RAC jurisdiction.

After I left RAC, our motion came to a vote and was defeated. I was unhappy with the outcome, not so much because my side had lost (as it always had done at times) but because of the reasons for the vote. I read the minutes of that RAC meeting carefully and was dismayed by the virtual irrelevance of most of the discussion to the purposes for which I thought RAC had been created. Scientific and lay members alike ignored the question of 'Will this make the world a safer place to live?' in favor of 'How will this play in Peoria?' The most influential opposition to our motion came not from traditional advocates of genetic regulation but from industrial and academic sources, such as the chairman of the Harvard Biosafety Committee. Their concern was that, if RAC withdrew regulatory supervision of most recombinant DNA research, political pressures for regulation at other levels would mount. From that time onward RAC became, in my opinion, a dummy committee with a sham function. Ostensibly it protects the public from danger, but in fact it serves more to protect Harvard from the Cambridge City Council and Genentech from the California State Assembly.

I believed then and believe now that RAC's action was shortsighted. Part of the background of recent court discussions in the recombinant DNA area is the implicit assumption that, if a technically sophisticated group such as RAC saw a need for Guidelines for so many years, they must have had a scientifically valid reason. Such is the legacy of political machinations.

2.2 CURRENT CONCERNS

How does all this relate to the issue of deliberate release of engineered microorganisms? Does my position on the NIH Guidelines imply that I favor complete *laissez-faire*? Not at all. Both deliberate release and large-scale production were among the few areas to which Guidelines would have continued to apply under the most liberal revisions ever proposed. Attempts by others to extend exemptions for small-scale laboratory experiments to these other levels have generally rested on devious logic which I cannot endorse.

For example, along with other RAC members, I labored long and ultimately successfully to exempt most cloning in *Escherichia coli* K12 from the Guidelines. First, K12 is a highly compromised laboratory strain, unable to compete with wild-type bacteria; second, we saw no reason to suspect that a small segment of foreign DNA would make it harmful. Both considerations apply with some force to the possible escape of a few bacteria from a small laboratory culture. If they escape, they are unlikely to spread; and even if they spread, no damage is likely to result. If we compare a culture of K12 produced by shotgun cloning of random samples of insect DNA to a culture containing a mixture of natural bacteria, such as an enrichment culture inoculated with soil or sewage, the former seems if anything safer than the latter. That would certainly be my decision if someone asked me which of the two cultures I would prefer to drink.

On the other hand, if someone were to isolate a single component strain from either culture intending to propagate and disseminate large amounts of it, I would consider it prudent to review its properties carefully for any unexpected problems it might cause. After the K12 exemption, some industrial representatives argued to RAC that, 'If something is safe in small amounts, it's safe in large amounts,' a position that, in context, impressed me as a deliberate misrepresentation. Small amounts of cultures of no known or predicted hazard do not warrant elaborate containment measures. That does not imply total absence of risk, and large-scale production is the stage at which more detailed checking is feasible.

2.3 NATURAL MICROORGANISMS COMPARED TO ENGINEERED ONES

As implied above, my concerns about large-scale propagation and dissemination are not restricted to engineered organisms but extend to natural ones as well. I see no rational scientific basis to distinguish between them. However, at least in the United States, regulation of industrial microorganisms has been spotty. Some areas may be overregulated, but others have been underregulated. For example, enzyme-containing detergents were eventually removed from the market-place, but after, rather than before, significant damage had been inflicted on both workers and consumers by these products of natural bacteria. The legal system makes it simpler to impose regulation on new processes rather than established ones, and complete equity may be attainable only in the very long range.

Much has been said and written about the novelty of engineered organisms, including the hypothesis that all imaginable heterologous gene combinations have already occurred in nature and have failed the test of natural selection. Discussion of this question can become confused by imprecision in the level of specification. It cannot be literally true that all conceivable individual gene combinations have been tried. The history of life on earth has been too short even to allow realization of all combinations that could result from reshuffling the human gene pool, let alone the vastly larger number of potential heterologous combinations that might be generated by interspecific transfer (Campbell, 1979; Regal, 1986).

One may of course reason that, since not all combinations have been tried, some particular untested combination might be selectively advantageous and therefore could spread and perhaps cause problems. The logic is unassailable, but in the context of evaluating engineered organisms, I think it misses the main point. The first thing that requires emphasis, in terms of the expectation of Western society in the 1990s is that, like it or not, we already inhabit an inherently uncertain world. A natural genetic lottery operates on all biological populations, including our own, and may generate disruptive new gene combinations at times and places beyond our power to predict. The uncertainties associated with engineered organisms must be measured against this background. The question is not whether individually novel combinations might be created, but whether molecular biologists are sampling a different (and perhaps more dangerous) pool of possibilities than nature samples on a grander scale. I see no reason to think that they do, but I am less certain of that answer than I am of my analysis of what question should be asked.

The expected impact of genetic novelties is closely related to one's position on evolutionary theory. All neo-Darwinian evolutionists see evolution as the joint result of mutation and selection, but only some imagine that the rate of evolution is limited by the rate of fresh (not necessarily unique) mutation (Nei, 1983).

CONCLUSIONS

Public discussion of the safety issues related to engineered organisms has frequently been clouded by decisions of regulatory bodies that were dictated more by politics than by logic. If political considerations are discounted, there are no convincing scientific grounds for distinguishing engineered organisms from natural ones. Because organisms of either type could pose unforeseen hazards, some safety testing is desirable before large-scale propagation.

REFERENCES

Campbell, A. (1979) Natural modes of genetic exchange and change. In: Morgan, J. and Whelan, W.J. (Eds.) *Recombinant DNA and Genetic Experimentation*, Pergamon Press, Oxford, pp. 21–7.

Nei, M. (1983) Genetic polymorphism and the role of mutation in evolution. In:

Nei, M. and Koehn, R.K. (Eds.) *Evolution of Genes and Proteins*, Sinauer Associates, Sunderland, Mass, pp. 165–90.
Regal, P.J. (1986) Models of genetically engineered organisms and their biological impact. In: Mooney, H.A. and Drake, J.A. (Eds.) *Ecological Studies*, Vol. 58, Springer-Verlag, New York, pp. 111–29.

Section 2

GENETIC TRANSFER AND THE ENVIRONMENT

Introduction of Genetically Modified Organisms into the Environment. Edited by Harold A. Mooney and Giorgio Bernardi

3 Impact of Human Civilization on Biological Evolution

WERNER ARBER
Department of Microbiology, Biozentrum, University of Basel, Switzerland

Essential strategies of genetic engineering are of direct evolutionary relevance. These include the transfer of genetic information between different kinds of organisms, site-directed mutagenesis, and the deliberate release of genetically altered organisms into the environment. The evaluation of risks inherent to these strategies should thus be based on an understanding of mechanisms acting in natural biological evolution, in particular with regard to spontaneous mutagenesis, selection processes, and isolation principles. These considerations can also reveal other impacts that the cultural evolution of mankind has on biological evolution. An appreciation of the relative importance of any single human intervention with ecological conditions in the biosphere can help us to put technological applications into a broad perspective with regard to their potential risks and to propose appropriate measures to prevent undesirable interference with the natural course of evolution.

3.1 A MULTITUDE OF DIFFERENT MECHANISMS CONTRIBUTES TO SPONTANEOUS MUTAGENESIS

For the purpose of our considerations we define a mutation as an alteration in the nucleotide sequence of a given genome. Such an alteration can be lethal and thus difficult to investigate experimentally, or it can alter a phenotypic trait and thus represent a mutation as defined in genetic experimentation, or else it can be silent. The latter type of genetic alteration can become functionally relevant at some later time, in conjunction with subsequent alterations of the genetic message. It is generally accepted that any living organism can tolerate only mutation rates affecting biological functions considerably less often than once per generation.

One source of mutation is the *limited fidelity of DNA replication*. These so-called replication errors group nucleotide substitutions, small deletions, small insertions, and other types of replication slippage products. Spontaneous

Introduction of Genetically Modified Organisms into the Environment. Edited by Harold A. Mooney and Giorgio Bernardi

mutation may also result from the intervention of *environmental mutagens* such as chemicals or radiation. Often these alter the chemical structure of one or a few nucleotides, or they can provoke breaks in the DNA chains.

Many living organisms possess elaborate *repair systems* which provide means to restore a majority of DNA sequence alterations resulting from incorrect replication and from the action of environmental mutagens.

Other important sources of spontaneous mutations are various, mostly *enzyme-mediated recombination processes*. In these reactions, existing nucleotide sequences are reassembled in a new order. Often these are internal rearrangements of DNA already forming part of the cellular genome, but sometimes invading external DNA is involved in recombinational mutation processes, e.g. upon integration of a viral genome into chromosomal DNA. Some of the recombinational mutations cause the duplication of certain DNA segments. On the other hand, these processes can also lead to deletion formation. Generally speaking, some recombination processes are conservative with regard to the number of nucleotides of the participating genomes, others not.

Systems of *general genetic recombination* are known to exist in many organisms. Their strategy is based on the recognition of extended nucleotide sequence homologies, which then serve for reciprocal recombination. These enzymes are known to function in some repair processes, e.g. where recombination between two damaged sister DNA molecules can serve for the restoration of an intact genome. General recombination helps to provide a wide genetic diversity upon sexual reproduction. It also directs recombination at homologous segments, e.g. transposable genetic elements, located at different sites of DNA molecules. These reactions of unequal crossing-over can give rise to DNA duplication, DNA inversion, and deletion formation. These processes are of immediate evolutionary relevance.

Transposition of mobile genetic elements is a recombination process, the biological significance of which seems largely to be evolutionary. In this process, DNA segments of defined length are translocated to other sites within the same or another DNA molecule. Depending on the transposable element involved, the criteria of target selection vary. Some transposable elements strictly select their transposition target on the basis of its nucleotide sequence. Others transpose more randomly into various sites. However, this does not necessarily mean that no selection is applied at all. A good example is IS2. This transposable element prefers to transpose into particular DNA segments, but within these segments, IS2 may insert into many different sites, which show no sequence homology among each other (Sengstag and Arber, 1987).

Well-known examples of *site-specific recombination* are the integration of bacteriophage genomes into the bacterial chromosome (Weisberg and Landy, 1983) and microbial DNA inversion systems (Plasterk and Van de Putte, 1984; Iida and Hiestand-Nauer, 1986, 1987). An essential part of the sites of recombination in such systems is determined by a consensus sequence. However, at decreasing frequencies, nucleotide sequences, which increasingly differ from the consensus, may still serve for recombination. Such recombination at secondary crossing-over sites open many possibilities to procure new DNA arrangements in microbial populations at low rates. Some of these rearrangements may represent new gene structures, which could be of utility in changing environmental conditions.

Still other, rare, recombination processes may be due to the action of enzymes, the principal role of which is to cut and to religate DNA strands, such as topoisomerases (e.g. DNA gyrase) and restriction endonucleases. How many of the so-called *illegitimate recombinants* are to be attributed to these processes or also to site-specific recombinations exerted on rarely used secondary sites remains to be seen.

In conclusion, a multitude of different mechanisms contributes to the formation of spontaneous mutations. Only some of these mechanisms can be considered to act fully at random in space and in time (Arber, 1984). Others might become activated under particular conditions, they might act specifically at particular locations on DNA molecules, or they might make use of randomness to a limited extent as part of a strategy serving to increase the chance of producing unique new DNA arrangements in order to provide a very wide genetic variability within populations of organisms.

3.2 LEAKINESS IN GENETIC ISOLATION PROVIDES MEANS FOR HORIZONTAL EVOLUTION

Efficient means of *reproductive isolation* inhibit the exchange of genomes between organisms that are not closely related. This principle is the basis of speciation and has certainly contributed to the genetic diversification.

Isolation barriers also limit the acceptance of small parts of a foreign genome, which might be spread around by natural gene vectors such as viruses or which might simply penetrate as free DNA molecules into a cell. Although invading DNA has to overcome a number of different barriers, such as the cell wall, restriction systems, lacking means to become established and to undergo propagation, the spontaneous acquisition of foreign genes has often been reported. It is particularly well documented with microorganisms.

In microbial systems, the following processes of horizontal gene flux are well known. In *transformation*, free DNA penetrates directly into a recipient cell. In *transduction*, genes from a donor cell become packaged into a viral particle, which then infects a recipient cell. In *conjugation*, a conjugative plasmid procures the efficient contact between two partner cells and it mediates the transfer of genes from the donor cell to the recipient cell.

In some of these processes *vector DNA molecules* form an essential part of the transfer system. This is the case in specialized transduction, where a viral genome serves as vector, and in conjugation, where a conjugative plasmid serves as vector. These vectors can become charged with small segments of the genome of the donor strain by some of the recombination processes already described above. These same processes can also help to stably establish a transferred gene segment in the infected recipient strain.

Because of isolation barriers, the successful transfer of genes between two different bacterial strains is relatively rare. The observed leakiness in the isolation appears to allow for a more efficient acceptance of small segments of foreign DNA than of larger DNA segments. The addition of single genes to a genome may also conceptually appear to be a better evolutionary strategy than an exchange of entire chromosomes. Virus-mediated horizontal transfer of one or a few genes has also been observed between several higher organisms. Hence we see in the horizontal, lateral gene transfer a perhaps general strategy of biological evolution.

While in *vertical evolution* entire genomes gradually undergo sequence alterations by various processes of spontaneous mutagenesis, in *horizontal evolution* the uptake of a small DNA segment, one or a few genes at once or even parts of a functional gene, can represent to the recipient cell an important acquisition, which might help it to cope with a new environmental constraint. In this view, the tree of biological evolution should be represented as a network, as represented in Figure 3.1.

3.3 SELECTION PLAYS AN UNCONTESTED ROLE IN BIOLOGICAL EVOLUTION

The world-wide spreading of drug resistance to enteric bacteria in the last few decades represents a convincing testimony for horizontal gene flux, and it also illustrates the potency of efficient selection. Molecular genetic analysis of antibiotic resistance traits, which had recently been acquired by enteric bacteria, has shown that these characters are often determined by genes carried on natural gene vectors and that many of them are also contained in transposable elements. There is no indication that either horizontal gene flux or transposition would be inducible by the presence of antibiotics. Therefore the wide use of antibiotics in human and in veterinary medicine in the last few decades must have influenced the widespread establishment of resistance genes by selection. Many of these genes were obviously acquired by horizontal transfer from other organisms. This is a rare event, but in the presence of antibiotics the very rare resistance recipient cells were heavily favoured and must have rapidly overgrown the non-resistant members of the microbial populations. Within 10 to 20 years of intensive use of antibiotics a high proportion of enteric bacteria have world-wide become resistant to one or often several antibiotics. This is a good measure

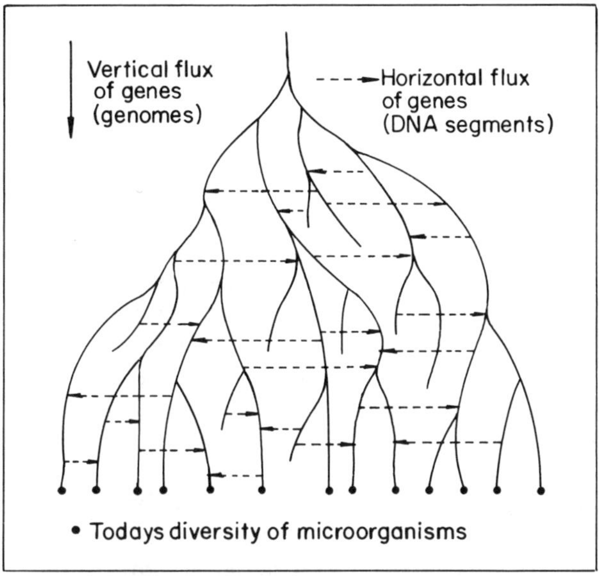

Figure 3.1 Microbial evolution represented as a network. As a function of time, genomes evolve by spontaneous mutagenesis as they are vertically transmitted from generation to generation. Relatively small segments of genomes are horizontally transmitted between different microbial species by any of a number of different mechanisms of interspecific gene transfer. This evolutionary tree with horizontal shunts between different branches should be regarded as multi-dimensional rather than two dimensional as drawn in this presentation.

for how fast potent selection can become evolutionary effective.

Evolutionary selection does not always act as immediately as in the application of antibiotics. Selective effects are normally based on more or less favourable living conditions. Under given conditions, some members of a mixed population of living organisms may propagate more readily than others, so that in the course of several generations the composition of the population may gradually change in favour of those organisms that find the living conditions most appropriate. The effect of selection becomes strongly manifest only in the course of time.

3.4 CULTURAL EVOLUTION MORE AND MORE INFLUENCES SELECTION AND THUS BIOLOGICAL EVOLUTION

Besides the described effect of the use of antibiotics many other human activities, as a matter of fact a great number of technological applications made with the motivation to serve for the benefit of mankind, may in the long term considerably change selection criteria in the biosphere and thus exert a direct influence on biological evolution. These effects of our civilization can be grouped into two categories.

The first category concerns the *management of land* for agriculture and human living. In these cultivated areas the original living conditions are drastically altered. Natural ecological equilibria are destroyed and often replaced by ecologically unstable plantations, frequently as monocultures 'protected' by appropriate interventions from competing weeds and from animals interfering with the growth of the crop. Under these artificially maintained conditions, natural biological evolution cannot function properly, so that these vast areas cannot any longer serve as breeding grounds for evolutionarily relevant genetic variants. As a result of this anthropogenic selection, many original inhabitants of these areas die out. The situation becomes particularly grave upon the transformation of forests into intensively used agricultural land.

The second category of influences of the civilization on the course of biological evolution acts through the increasing *pollution of soil, air, and water*. Pollution is also spread through the air and through the water to those areas of the globe where human activities have not yet exerted a direct influence on the living conditions. These conditions suffer from deterioration under the increasing concentrations of pollutants. Therefore evolutionary relevant selection also undergoes changes in the so-called wilderness, which includes virgin forests, high mountain regions, deserts, and also the oceans. Practically no region of the planet is protected from such effects. Again one must expect a drastic reduction in biological diversity as a result of these interferences at the level of evolutionary selection.

Both from the massive transformation of land for direct human use as well as from the influence of increasing concentrations of pollutants including carbon dioxide one expects considerable world-wide climatic changes. This can again affect efficiently the selection processes acting in biological evolution.

3.5 MAINTENANCE OF A RICH BIOLOGICAL DIVERSITY IS ESSENTIAL BOTH FOR FUTURE BIOLOGICAL EVOLUTION AND FOR FUTURE CULTURAL EVOLUTION

Mainly for the reasons discussed above, mankind is in the process of involuntarily reducing genetic diversity. Not only are many living species dying out but others, particularly those of agricultural importance, steadily lose their diversity of genetic variations as a result of present-day agricultural practices. In the context of the issues discussed here, two strong reasons should motivate us to call for urgent measures to preserve the rich genetic diversity:

1. As we have outlined above, genetic rearrangements and in particular horizontal gene flux are important sources for spontaneous mutations, a basic condition for living organisms to cope with changing environmental conditions. Any reduction in genetic diversity reduces the evolutionary possibilities.
2. Biotechnological applications are likely to widely influence the life of human societies of the future. Many gene products may serve in the future as biologically active products in medicine and in many other fields of human activities. It is impossible to predict today which of these applications will be the most beneficial. Nor do we know in which living organisms potentially useful genes are. With every species that dies out genes for interesting products could become lost for ever. Genes should be considered as unique resources for the human civilization of the future. A rich genetic diversity should also be maintained for these reasons.

In summary, it is our responsibility to safeguard the genetic richness as a basis for normal biological evolution and for the further progress of our civilization.

3.6 THE SPONTANEOUS APPEARANCE OF A HIGHLY SPECIFIC, NEW NUCLEOTIDE SEQUENCE IS STATISTICALLY EXTREMELY RARE

The nucleotide sequence of a gene of average size is about 1000 bp. This sequence is built by a specific linear arrangement of 4 different nucleotides. Therefore, one expects 4^{1000}, that is 10^{602}, different specific 1000-bp sequences to be possible.

Based on an estimation of the volume of the biosphere of our planet and on our knowledge of the volumes of living cells, one can estimate the biosphere to contain roughly 10^{30} living cells. Life exists on earth for 3×10^9 years, that is 10^{17} seconds. Assuming that each living cell would try out one new nucleotide arrangement every second, which would correspond to an excessively high mutation rate, only about 10^{47} of the 10^{602} possibilities of specific gene structures could have been tested by today.

Although it is clear that many slightly different nucleotide sequences can code for the same biological function, this simple calculation does suggest that the spontaneous generation of a particular biological function might be a rare event. For this reason, the preservation of existing genes and thus of a high genetic diversity can prevent us from the permanent loss of existing biological functions that could become of increasing importance at some future time. It is unlikely that the gene for a particular biological function can be generated *de novo* any time that there is a need for it.

3.7 CONCLUSIONS WITH RESPECT TO THE RELEASE OF GENETICALLY DESIGNED ORGANISMS IN THE ENVIRONMENT

3.7.1 *IN VITRO* RECOMBINATION BYPASSES ISOLATION BARRIERS

One of the strategies of gene technology implies the horizontal transfer of a gene into a cell of another species of organisms, e.g. a mouse gene into a bacterial cell. As we have discussed, horizontal gene transfer is an important principle acting in natural biological evolution. However, relatively little is yet known about these phenomena, particularly on the range of organisms into which genes usually become transferred horizontally. It is clear, however, that in *in vitro* recombination, existing isolation barriers are often bypassed. In other words, the natural leakiness of isolation barriers becomes increased in the *in vitro* gene transfer. Biological evolution is deliberately guided into a chosen direction. Although biological evolution could go into the same direction also by natural means, the experimental approach brings an increase in the rate of such occurrences. The evolutionary appearance of a new gene combination is not any longer a matter of chance; it is the reflected goal of the project.

3.7.2 SITE-DIRECTED MUTAGENESIS ALSO INFLUENCES EVOLUTION DELIBERATELY

Site-directed mutagenesis serves on the one hand to explore gene functions systematically and on the other hand to specifically alter, e.g. improve, gene functions, once their mechanisms have been clarified. Again, any such genetic alteration could also occur spontaneously, but the probability of most specific genetic alterations is low. Hence genetic experimentation also influences evolution at the level of mutation, and this again into a designed direction. Alteration of the genetic message can either alter the specificity of the gene product or the efficiency of gene expression. Both of these influences can represent evolutionary steps.

3.7.3 GENETIC EXPERIMENTATION CAN BYPASS PRIMARY SELECTION

In genetic experimentation, molecular as well as classical, large clones of new genetic combinations can be produced in the absence of naturally acting selection principles. This can bypass the need to establish founder colonies, an often stringent condition which stands in the way to the success of a new trait in ecological populations.

3.7.4 *IN VITRO* RECOMBINED GENES CAN LATER BECOME TRANSFERRED HORIZONTALLY UNDER NATURAL CONDITIONS

There is no reason not to assume that cloned genes as well as any other gene could, at one time or another, become transferred horizontally to other organisms forming part of the ecological population in which the host of the cloned gene might live after its release into the environment. At this level, natural isolation barriers are expected to work properly and display their normal leakiness.

3.7.5 FURTHER POINTS TO BE CONSIDERED FOR A WISE, RESPONSIBLE APPLICATION OF RECOMBINANT DNA STRATEGIES

With respect to biological evolution, recombinant DNA techniques hardly represent entirely new aspects. Leakiness of genetic isolation, spontaneous mutagenesis, and mechanisms to bypass selective forces are known to exist under natural conditions. However, genetic experimentation, whether molecular or classical, can interfere with natural evolution (1) by the design of the experiment which can push evolution into a chosen direction and (2) by the possibility of introducing altered organisms at high concentrations into the environment. Particular attention should be given to these aspects. In view of the discussed importance of maintaining a high genetic diversity in our biosphere, any experimental design should avoid causing directly or indirectly a reduction in the genetic richness. For example, if a new trait or a new property is to be introduced into a crop of agricultural importance, care should be taken not to use a single clone. Rather, the gene for the particular trait should be introduced into a large number of different natural varieties of the crop, a measure which could help to preserve the richness of the natural gene pool of the concerned species. This kind of strategy may not be in the immediate interest of those propagating commercial applications. However, a wide scientific consent of their general importance could help such measures to be followed.

For a better evaluation of the long-term effects that the introduction of genetically designed organisms into the environment can have, a better understanding of biological evolution is an urgent need. Research in these fields deserves particular attention. Its results can also help us in the appreciation of any of the many different other influences of human civilization on biological evolution, such as those outlined above. Release of genetically modified organisms, if it respects a high genetic diversity, is likely to interfere with natural biological evolution to a much lesser extent than many other human activities, particularly if these alter the living conditions and thus evolutionary active selection.

The principle that any functional gene represents a unique entity of imminent importance for future biological as well as cultural evolution deserves a wide recognition by the scientific community. This could also help its acceptance by the general public. This is a prerequisite for politically effective measures to be taken in view of diverting the imminent threat that genetic diversity becomes drastically reduced within the next few human generations. Recombinant DNA techniques are not the main cause of this danger; rather, if applied with responsibility and wisdom, it is likely to serve mankind to remedy the present ecological abuses and thus to preserve a high genetic diversity.

REFERENCES

Arber, W. (1984) Natural mechanisms of microbial evolution. In: Arber, W., Illmensee, K., Peacock, W.J. and Starlinger, P. (Eds.) *Genetic Manipulation: Impact on Man and Society*, The ICSU Press and Cambridge University Press, Cambridge, pp. 1–14.

Iida, S. and Hiestand-Nauer, R. (1986) Localized conversion at the crossover sequences in the site-specific DNA inversion system of bacteriophage P1. *Cell* **45**, 71–9.

Iida, S. and Hiestand-Nauer, R. (1987) Role of the central dinucleotide at the crossover sites for the selection of quasi sites in DNA inversion mediated by the site-specific Cin recombinase of phage P1. *Mol. Gen. Genet.* **208**, 464–8.

Plasterk, R.H.A. and Van de Putte, P. (1984) Genetic switches by DNA inversions in prokaryotes. *Biochim, Biophys. Acta* **782**, 111–19.

Sengstag, C. and Arber, W. (1987) A cloned DNA fragment from bacteriophage P1 enhances IS2 insertion. *Mol. Gen. Genet.* **206**, 344–51.

Weisberg, R.A. and Landy, A. (1983) Site-specific recombination in phage lambda. In: Hendrix, R.W., Roberts, J.W., Stahl. F.W. and Weisberg, R.A. (Eds.) *Lambda*, Vol. II, Cold Spring Harbor Laboratory, Cold Spring Harbor, pp. 211–50.

4 Epistatic and Pleiotropic Effects on Genetic Manipulation

ALLAN CAMPBELL
Department of Biological Sciences, Stanford University, Stanford, California, USA

Most biologists agree that there there is no special hazard associated with engineered organisms as a group. Most of us also agree that one could (if one wished) design organisms expected to cause problems if released, at least at the local level. The major remaining doubts therefore focus on the possibility of inadvertently constructing something that is undesirable for reasons that are unanticipated. Epistasis and pleiotropy cover some interactions between genes or their products that are more complex (and therefore sometimes harder to predict) than cases where independent unit functions are simply added to or subtracted from genomes. Both terms were invented to describe mutations that are found in natural (nonengineered) populations.

The term *epistasis* refers to situations where the phenotypic consequences of mutation at one locus are obscured by the state of a second genetic locus in the same cell. One relevant bacterial example is the determination of antibiotic resistance. Streptomycin-resistant mutations in *Escherichia coli* affect the target of streptomycin action, a ribosomal protein; in bacteria carrying a duplication of the chromosomal gene that encodes this protein, sensitivity is dominant over resistance, because a cell in which 50% of the target protein is inactivated does not survive. *E. coli* cells can also become streptomycin resistant through acquisition of plasmids encoding enzymes that destroy the antibiotic. In a cell carrying such a plasmid, the state of the chromosomal locus has no effect on response to streptomycin. The plasmid gene is epistatic to the chromosomal locus.

A *pleiotropic* mutation is a single genetic change that affects more than one phenotypic trait. There are many bases for pleiotropy. A single protein or a single product may function in multiple pathways; thus mutations of *E. coli* that inactivate enzymes in the pathway for molybdopterin synthesis cause simultaneously inability to reduce nitrate and some organic sulfoxides, inability to oxidize trimethylamine, and resistance to added chlorate because nitrate reduction, trimethylamine oxidation, and chlorate sensitivity all

Introduction of Genetically Modified Organisms into the Environment. Edited by Harold A. Mooney and Giorgio Bernardi

depend on enzymes that use a common molybdopterin cofactor. Missense mutations in proteins that function in central biosynthetic processes can affect the fidelity of macromolecular synthesis and generate a host of incidental changes, as seen in mutations to streptomycin resistance (protein synthesis) and rifampicin resistance (RNA synthesis). The multiple effects of mutations to streptomycin dependence served as one of the first clues to the nature of the affected protein, before it was known to be ribosomal (Spotts and Stanier, 1961). Mutations affecting components of global regulatory systems can simultaneously affect expression of all the genes controlled by that system. Thus, mutants unable to make cyclic AMP fail to turn on many separate genes for fermentation of different sugars, while *recA* mutations reduce genetic recombination and expression of the SOS repair system as well as spontaneous production of phage by lysogenic bacteria. Impairment of critical steps in the developmental program of multicellular organisms can create bizarre monsters such as homeotic mutants in which whole organs are misplaced.

Biologists have little difficulty incorporating pleiotropic mutations into the conceptual framework of cellular and organismal determination. Both metabolism and development are affected by complex arrays of interacting components. The effects of individual genetic changes are frequently an incisive tool for detecting those interactions (Botstein and Maurer, 1982). On occasions (as in the early days of biochemical genetics), geneticists have deliberately avoided pleiotropic mutations because they complicated the straightforward analysis of gene action. This has been the shortest course to determining basic mechanisms, but has sometimes created a textbook genetics highly biased toward mutations and genes with simple effects.

Some developmental mutants have such dramatic effects on organismal form that they force reflection on the determinative role of the genes that have mutated. If changing a single gene causes a leg to project from a fly's head, obviously that gene cannot, by itself, specify all the information required for leg development. The same distinction between induction and specification arises in experimental embryology. Tissues can be induced to differentiate along certain lines by natural biological signals, such as cellular contact or gradients of morphogens, but also sometimes by very simple chemical or physical stimuli, such as ion concentration, pH, or even contact with glass beads. Clearly the information for the developmental pathway is specified not by the inducing signal but by the induced tissue (ultimately, by its genome). The inducing stimulus selects one pathway from a limited number of alternatives prespecified by that genome. Likewise, a gene whose alteration redirects development into a different pathway cannot be the major source of information for that pathway, but must control some switch that activates many other genes.

In engineered organisms, we may ask whether foreign genes exert epistatic or pleiotropic effects and, if so, what those effects might be. Usually, the effects will be deleterious rather than advantageous to the organism in question and therefore limit its spread, but there might be exceptions. In principle, answers might either come from experiment (based on surveying the engineered organisms isolated to date) or from theory. As to experiment, I know of no facts to indicate that genetic manipulation has produced any radically surprising phenotypes. On the theoretical side, we may at least try to define the issue. In the early days of recombinant DNA research, some champions of the technology emphasized that an *E. coli* cell harboring a small segment of exogenous DNA is, after all, still *E. coli*; that little piece of exogenous DNA could hardly change it into a totally new organism. Our discussion of pleiotropy indicates the difficulty with literal acceptance of that position. If a very small change in the *E. coli* genome can effect multiple alterations in phenotype, one can hardly doubt that introduction of some foreign genes would have similar drastic effects.

Nevertheless, there is some justification for emphasizing the small size of the added piece. Even if it does generate a radical change in phenotype, it cannot do so by adding large amounts of new information; rather, like known pleiotropic mutations, it must mainly select prespecified but unexpressed potentialities of the resident genome. In that respect, I would expect that most such effects would not be far removed from changes that can occur by mutation in the nonengineered host.

One highly visible effect of genetic manipulation has been the development of giant mice by deliberate amplification of growth hormone genes (Palmiter *et al.*, 1983). Any fantasies of the earth being overrun by giant rodents escaped from the laboratory are dispelled by knowledge that gigantism and dwarfism are traits that arise in all mammals by spontaneous mutation (perhaps sometimes by gene amplification) and that have not won out in natural selection. We should expect that introduction of heterologous genes for developmental hormones would sometimes generate bizarre monstrosities of unpredicted phenotype, but we should also expect similar abnormalities to arise by spontaneous mutation.

4.1 MICROEVOLUTION VERSUS MACROEVOLUTION

The importance of pleiotropy spills over into a question about natural evolution that has long been debated. For most of the present century, the dominant figures of population genetics have belonged to the microevolutionary school, which attributes most evolution to the gradual accumulation of small genetic changes that collectively produce major alterations in phenotype. Throughout the same period, a minority group, but one that has included some geneticists of great distinction, has proposed that sudden large discrete changes in genome organization are the pivotal evolutionary events. The most explicit advocate of macroevolution was Richard Goldschmidt, but some of the ideas propounded more recently by Barbara McClintock, as well as the recent vogue of the 'punctuated equilibrium' concept of paleontologists, bear some of the same stamp (Maynard Smith, 1983).

Personally, I have found many discussions of this subject hard to follow because the issues are seldom explicitly defined. When microevolutionists speak of small changes, do they mean small changes in DNA sequence (such as simple base substitutions compared to complex rearrangements) or do they mean changes with small phenotypic effects (such as typical missense mutations or amplifications of genes already present in multiple copies), as contrasted to highly pleiotropic mutations of the types discussed here? The distinction was probably harder to make in the premolecular era, but it seems absolutely central. To my mind, the most important distinction is between simple and complex changes in DNA sequence or, more precisely, between changes that occur readily (which includes not only replication errors but also rearrangements facilitated by specific recombinases and transposases) contrasted to very rare or multiple events. Changes in the former category, however substantial their phenotypic effects, are happening all the time and therefore are already represented in the gene pools of existing species at frequencies set by rates of mutations and selection and parameters of population structure. For that reason, they seem readily assimilable into the general framework of microevolutionary theory. On the other hand, if DNA changes sufficiently rare to be effectively unique are major evolutionary determinants, a truly different, far more stochastic, description of evolution is required.

Macroevolution (especially in the sense of rare DNA changes indicated here) remains a minority viewpoint at best, for reasons I will not detail here. As the introduction of a foreign DNA segment constitutes a genuine macromutation, the possibility that such events are important in natural evolution is at least worth considering.

4.2 CONCLUSIONS

Small changes in genomes, either by mutation of critical components or by artificial manipulation, can effect major alterations in phenotype. However, even when the changes create new forms with a high degree of organization, the principal determinants of form lie in the residual genome rather than at the site of genetic alteration. The altered elements act mainly to elicit unexpressed potentialities of the genome. Two consequences are relevant to engineered organisms. First, engineering small alterations in genotype may have large (sometimes unpredicted) phenotypic consequences; second, equivalent changes are likely to have occurred also in natural populations without genetic manipulation. Assessment of the possible impact of engineered organisms depends on the extent to which evolutionary rates are limited by mutation rates and on the relative importance of microevolution versus macroevolution.

REFERENCES

Botstein, D. and Maurer, R. (1982) Genetic approaches to the analysis of microbial development. *Ann. Rev. Genet.* **16**, 61–84.

Maynard Smith, J. (1983) The genetics of stasis and punctuation. *Ann. Rev. Genet.* **17**, 11–25.

Palmiter, R.D., Norstedt, G., Gelinas, R.E., Hammer, R.E. and Brinster, R.L. (1983) Metallothionein-human GH fusion genes stimulate growth of mice. *Science* **222**, 809–14.

Spotts, C.R. and Stanier, R.Y. (1961) Mechanism of streptomycin action on bacteria: a unitary hypothesis. *Nature* **192**, 633–7.

5 Modulation of Replication Efficiency by Non-coding Sequences and by the Environment

GIORGIO BERNARDI, REGINA GOURSOT, EDDA RAYKO, RENÉ GOURSOT, BAYA CHERIF-ZAHAR, AND ROBERTA MELIS
Laboratoire de Génétique Moléculaire, Institut Jacques Monod, Paris, France

5.1 INTRODUCTION

Two recent observations from our laboratory are relevant to the subject of this volume. They both concern the modulation of replication of the mitochondrial genome from cytoplasmic petite mutants of yeast. In the first case, modulation is due to non-coding sequences flanking the *ori* sequences, in the second to an environmental factor, temperature.

The underlying questions are of a very fundamental nature. In the first case, the problem is that of the influence of sequences flanking a given DNA segment on the functional properties of the latter. Obvious effects can be expected, and are found, owing to the presence or absence of regulatory signals in the flanking sequences. A more interesting situation arises when this is not the case and the effects are presumably just due to changes in DNA structure around the sequence under consideration. This is exactly the case investigated by us, by looking at the influence of non-coding sequences that flank the *ori* sequences of the mitochondrial genome of yeast.

The second question concerns the effect of the environment on the genome. In evolution, the environment mainly acts as a selection agent; organisms adapt to diverse environments by being selected for. The environment can also cause, in a direct or an indirect way, chemical alterations in the genome, as exemplified by the actions of mutagens. Another way in which the environment can affect the genome is by directly

Introduction of Genetically Modified Organisms into the Environment. Edited by Harold A. Mooney and Giorgio Bernardi

inducing changes in its higher-order structure, as shown in the results to be described here.

5.2 THE MITOCHONDRIAL GENOME OF YEAST

The mitochondrial genome of wild-type, respiratory competent *Saccharomyces cerevisiae* is made up of 25–50 genome units (in haploid and diploid cells, respectively), which segregate into the buds at cell replication. Each one of the identical genome units (Figure 5.1) comprises genes encoding ribosomal RNAs and transfer RNAs, as well as genes encoding protein subunits of enzyme complexes associated with respiration and energy production. The latter genes contain, in a number of strains, intervening sequences, some of which contain open reading frames. All coding sequences only represent one third of the mitochondrial genome, the rest being formed by non-coding, mostly intergenic, sequences. The latter are formed by long AT spacers, only 5% in GC, and by about 200 GC clusters, about 60% in GC (de Zamaroczy and Bernardi, 1985, 1986a, 1986b, 1987).

As intergenic sequences are highly repetitive, they confer an extreme instability to the mitochondrial genome of yeast, which is dispensable (yeast being able to survive through fermentation). Indeed, excisions of DNA segments from the mitochondrial genome occur very frequently at pairs of direct repeats essentially located in intergenic sequences (de Zamaroczy *et al.*, 1983). The excised segment is amplified in tandem to become the repeat unit of the defective genome of the cytoplasmic respiratory deficient petite mutants. These defective genomes segregate into the buds and may undergo secondary excisions (Figure 5.2). Replication of spontaneous petite mutants is usually insured by the presence in its repeat units of one of the three to four functional *ori* sequences (de Zamaroczy *et al.*, 1984; Baldacci *et al.*, 1984) from the wild-type genome (Figure 5.3). Such *ori* sequences are about 300 bp long and comprise several features : (1) two GC clusters, A and B, which can form with the intervening AT sequences an A–B fold (see Figure 5.5); (2) a long AT sequence *l*; and (3) a GC cluster, C, which is followed by a region *r* in which mitochondrial RNA polymerase starts the synthesis of the RNA primer of one nascent DNA chain; the RNA primer for the other nascent DNA chain is started at the other side of cluster C (Figure 5.3).

5.3 THE EFFECT OF DNA REGIONS FLANKING THE ORI SEQUENCE ON REPLICATION EFFICIENCY

If one tests the replicative efficiency of *ori+* petites (namely of petites carrying an intact *ori* sequence in their mitochondrial genome) by crossing

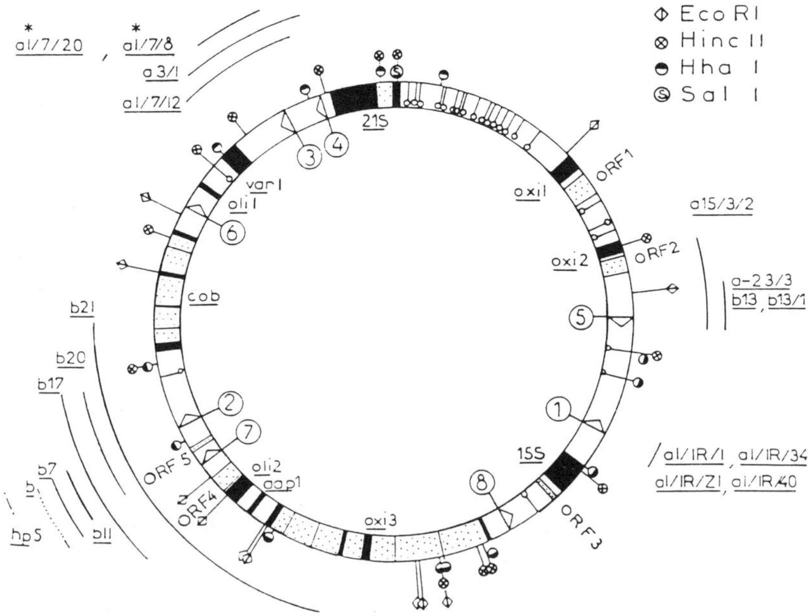

Figure 5.1 Physical map of a long mitochondrial genome unit of wild-type *S. cerevisiae*. Black areas correspond to mitochondrial genes or their exons (the last exon block of *oxi*3 comprises exons A 6–8 and introns a1 7,8); dotted areas to intervening sequences and intergenic open reading frames (ORF1–ORF5; Colin *et al.*, 1985); radial lines indicate tRNA genes (the thr1 gene is the only one to have an anticlockwise orientation). White areas correspond to long AT spacers with embedded short GC clusters. Among mitochondrial genes, *oxi*1, 2, and 3 encode subunits II, III, and I, respectively, of cytochrome c oxidase; *cob*, cytochrome b; *aap*1, *oli*2, and *oli*1, subunits 8,6, and 9 of ATPase; *var*1, a protein associated with the small mitochondrial ribosome subunit; 9S corresponds to the central part of tRNA synthesis locus; 15S and 21S are the genes for the small and large ribosomal RNAs, respectively. Triangles indicate the location of *ori* sequences 1–8; they point in the direction cluster C to cluster A.

The localization of the repeat units from the mitochondrial genomes of petite mutants used in the present work on the physical map of the mitochondrial genome of wt cells is shown. When several petite genomes correspond to the same localization, they are of about equal sizes. In the case of petite b, the dotted line corresponds to a central deletion; in that of petite hp5, excision sequences were not determined. Some restriction sites are indicated.
(From Rayko *et al.*, 1988.)

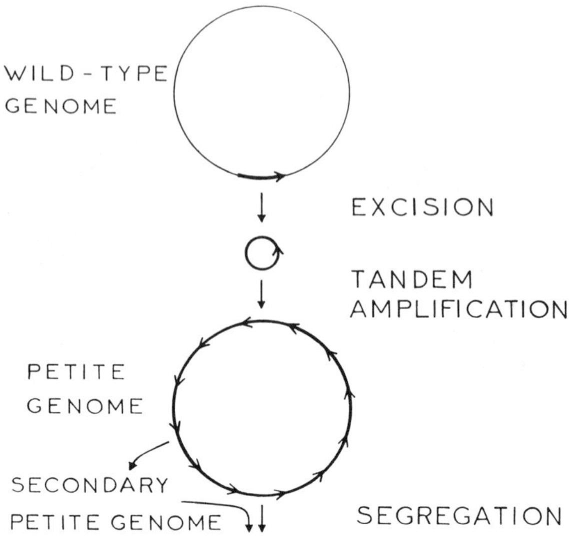

Figure 5.2 Scheme of the mechanism of formation of the defective mitochondrial genomes from spontaneous petite mutants. A mitochondrial genome unit from wild-type cells undergoes an excision. The excised segment is tandemly amplified to form a mitochondrial genome unit of the petite mutant. This genome replicates and segregates into the buds. It may also undergo secondary excisions (from Bernardi, 1983).

with either a wild-type strain (suppressivity test) or another petite, the general rule is that the mitochondrial genome made up of shorter repeat units (and containing therefore more *ori* sequences) will show a higher replicative efficiency, as shown by its predominance in the progeny. We have found, however, exceptions to this rule (Figure 5.4). Indeed, some petite genomes characterized by longer repeat units may replicate better than other ones carrying shorter repeat units. Since comparisons were made among petite genomes carrying the same *ori* sequences, the only difference was the presence or absence of a few dozens of nucleotides in the regions flanking the *ori* sequences. Such regions are made up by AT spacers and GC clusters, may be located upstream or downstream of *ori* sequences, are different in length and primary structure, and therefore unlikely to carry any specific sequence signal.

The conclusion of these investigations (Rayko *et al.*, 1988) is that non-coding sequences flanking the *ori* sequences may affect the efficiency of replication in all likelihood by altering the higher-order structures of the mitochondrial genome. This conclusion appears to be of general interest, since it does not seem to be limited to mitochondrial *ori* sequences of yeast. Indeed, the deletions in a 500 bp region of plasmid PT181 (a 4.4 Kb

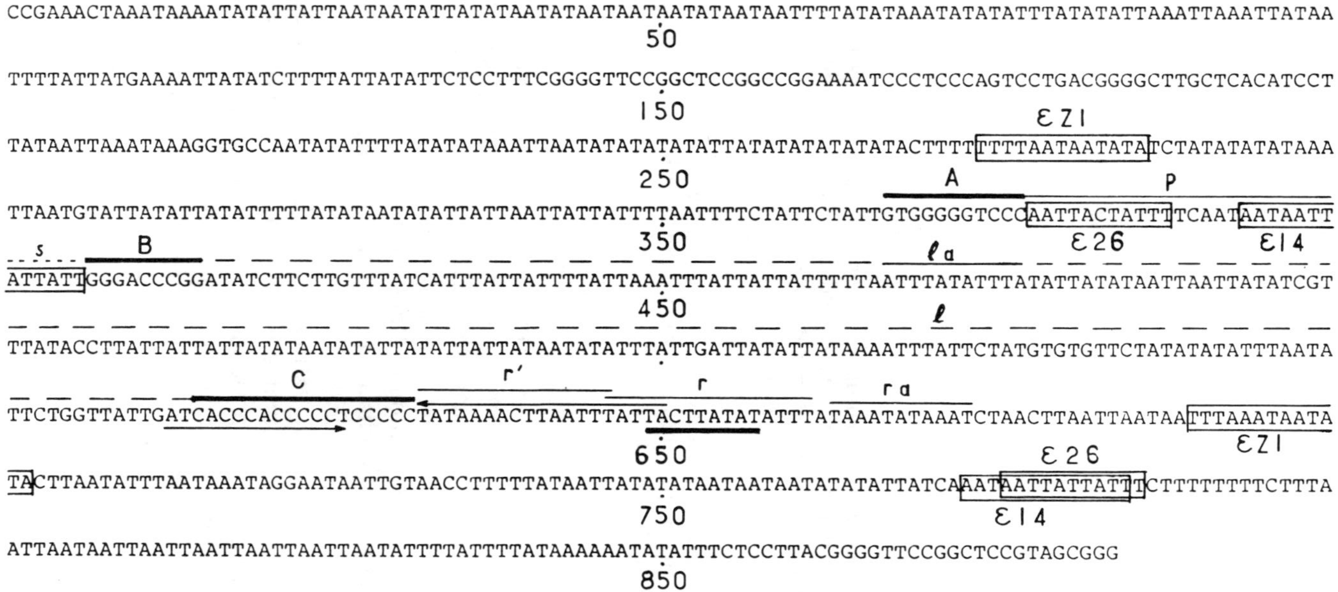

Figure 5.3 The repeat unit of the mitochondrial genome of primary petite a1/1R/1 (see Gaillard *et al.*, 1980) is 884 bp in size and contains an *ori* sequence, *ori*1. This comprises three GC clusters, A, B and C (heavy overlines). Cluster B is separated from cluster A by two short AT sequences *p* and *s*, and from cluster C by a long AT sequence *l* (broken overline). Cluster C is followed by sequence *r* (light overline) which comprises, on the strand shown, a nonanucleotide (heavy underline) where transcription of the RNA primer of one nascent DNA chain starts (arrow pointing to the left); transcription of the RNA primer for the other nascent DNA chain starts before cluster C (arrow pointing to the right) (Baldacci *et al.*, 1984). Sequence *r* is followed by a sequence *r*a which might pair with sequence *l*a (within the *l* sequence) to form a hypothetical tertiary structure (de Zamaroczy *et al.*, 1984). In the case of *ori*1, sequence *r* is preceded by a duplication, sequence *r*′. Excision sequences (ε) of the secondary mitochondrial genomes from petites 14, 26, and Z1 (see Gaillard *et al.*, 1980; de Zamaroczy *et al.*, 1984) are indicated by boxes (From Goursot *et al.*, 1988).

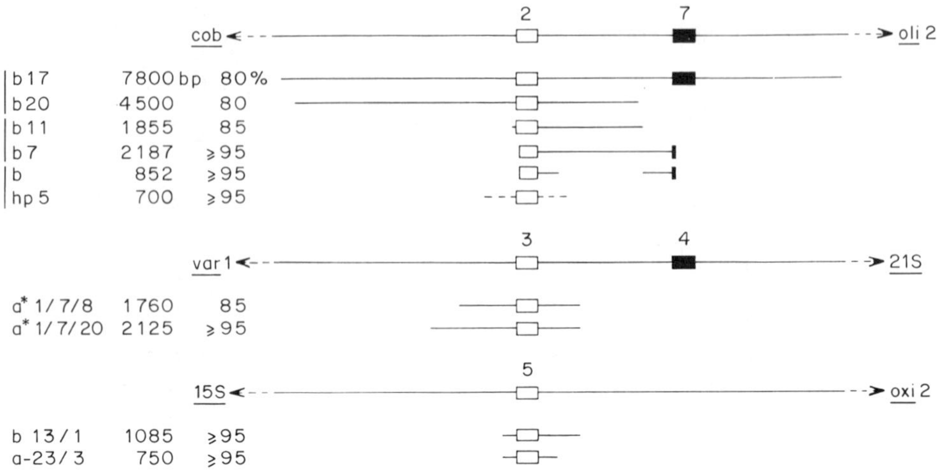

Figure 5.4 Repeat units of petites harbouring mitichondrial genomes carrying identical *ori* sequences and exhibiting suppressivities and/or competitive abilities in petite × petite crosses which do not follow the repeat unit size rule (see Discussion). Repeat unit sizes and suppressivities are indicated. Open boxes indicate active *ori* sequences 2,3, and 5, blackened boxes indicate inactive *ori* sequences 4 and 7 (de Zamaroczy *et al*, 1984). The *ori* sequences are all oriented with cluster C to the right of cluster A. Flanking genetic markers are indicated (see also Figure 5.1). Vertical bars on the left indicate the pairs of petite repeat units which are compared and discussed. (From Rayko *et al.*, 1988.)

multicopy plasmid of *Staphylococcus aureus*) external to the minimal replicon decreased the ability of the plasmid to compete with a coexisting incompatible plasmid, whereas in the homoplasmic state the deletions affected neither copy number nor plasmid stability (Gennaro and Novick, 1986). These deletions appeared to affect the interaction of RepC, a trans-acting initiator protein which is rate limiting for replication, and the plasmid origin of replication.

Concerning the mechanism by which the extra sequences favour replication, it should be pointed out that in different cases the extra sequences are different in primary structure, size, and position relative to the *ori* sequences (Figure 5.4). This makes it unlikely that replication is favoured by specific 'replication enhancement' sequences, and rather suggests that the effect is due to differences in the DNA or nucleoid (see, for instance, Rickwood *et al.*, 1981) structure in the different petite genomes under consideration. Such differences can be visualized as differences in DNA bending, DNA superhelicity, and/or DNA–protein interactions which may modify the secondary and tertiary structure of the *ori* sequences themselves (see de Zamaroczy *et al.*, 1984) and, consequently, the initiation of DNA replication; obviously effects on the elongation rate of newly synthesized DNA strands are also conceivable.

Since the flanking sequences of *ori* sequences in the petite genomes under consideration are made up of AT spacers and GC clusters, our results provide evidence that intergenic non-coding sequences play a role in the modulation of an essential genome function, such as replication, and substantiate the general idea, supported by other lines of evidence (Bernardi, 1982, 1983; Bernardi and Bernardi, 1986; de Zamaroczy and Bernardi, 1986a, 1987), that non-coding sequences do play a physiological role in genome function.

As a final remark, it should be noted that effects similar to those described here might be responsible for at least some of the cases in which recombinant plasmids carrying particular inserts fail to replicate.

5.4 THE EFFECT OF TEMPERATURE ON REPLICATION EFFICIENCY

The second observation concerns the effects of temperature on the efficiency of replication in petite genomes that have a thermosensitive structure in their *ori* sequences. Two petites from our collection, a1/1R/14 and a1/1R/26 (called 14 and 26 henceforth), are made up of repeat units containing *ori* sequences which are partially deleted in their cluster A and also in the neighbouring sequences. As a consequence, these *ori* sequences lack the A–B fold which requires an intact A cluster. They can form, however, replacement folds which are only made by AT base pairs (Figure 5.5). The lower thermal stability of the replacement fold compared to the A–B fold suggests that temperatures within a physiological range for yeast may affect its structure. We have, therefore, tested the suppressivity of these petites and compared them with that of a closely related petite which comprised, however, an intact *ori* sequence, a1/1R/Z1 (or Z1). As shown in Table 5.1, the petites containing the defective *ori* sequences showed a strong temperature dependence (in the 23–33°C range) of its replication efficiency as tested by suppressivity, whereas the control petite showed essentially no change under the same conditions.

The results presented (Goursot *et al.*, 1988) indicate that an environmental factor, temperature, can reversibly affect the replicative ability of a genome by altering its secondary (and, possibly, its tertiary) structure. Indeed, (1) these changes cannot be ascribed to enzymes involved in DNA replication, as in temperature-sensitive mutants, since the petites discussed here are isonuclear and lack mitochondrial protein synthesis, like all petites; (2) the different effects of temperature on the replicative ability of petites Z1, 14, and 26 show an excellent correlation with those expected from the secondary structures of the postulated A–B fold and replacement folds (de Zamaroczy

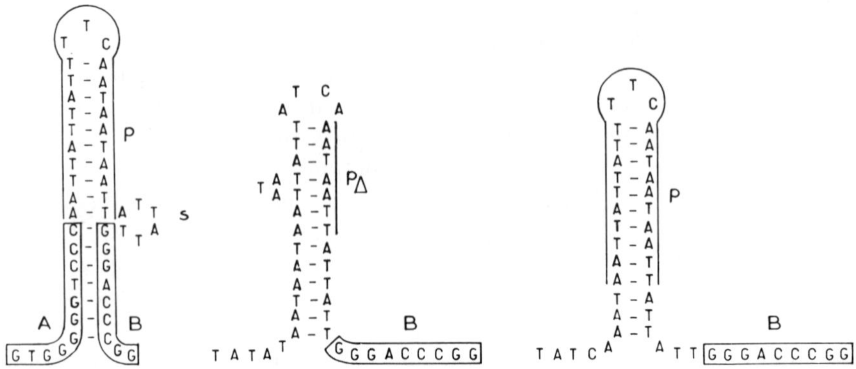

Figure 5.5 Potential secondary structure of the 'A–B fold' of the *ori* sequence present in the repeat units of mitochondrial genomes from petites a1/1R/1 and Z1 (*ori*1) and of the 'replacement folds' that can be formed in the *ori*1 sequence present in the mitochondrial genomes of petites 14 and 26. In the case of petite 14, the residual nucleotides from the partially deleted *p* stretch (*p*Δ) can generate a hairpin structure within nucleotides from the preceding repeat unit, but the stem, only formed by 13 AT nucleotides, carries different terminal and side loops compared to the A–B fold. In the case of petite 26, the upper part of the stem and terminal loop are identical to those of the A–B fold, but the lower part is replaced by three A:T pairs (three nucleotides are derived from the preceding repeat unit). The sequences involved in the structure shown (GC clusters A and B, sequences *p* and *s*) are those indicated in Figure 5.1. (From Goursot *et al.*, 1988.)

Table 5.1 Suppressivity of petites 14, 26, and Z1 at different temperatures

Temperature (°C)[a]	Suppressivity of petites (%)[b]		
	14	26	Z1
23	90	89	99
28	79	85	99
33	45	58	97

[a]Temperatures indicated were applied to precultures and to cultures used in crossings and in the incubation of diploids. Precultures in stationary phase were diluted ten times upon setting up the cultures used in crossings.
[b]Suppressivity was tested by crossing petite strains (MATa, adel) against the grand strain B (MATa, trpl, hisl). The percentage of petite/total diploid colonies was determined by counting about 1000 colonies on minimal agar.

et al., 1984), an effect on tertiary structures being also possible.

Interestingly, the conclusion that DNA secondary structure is required for *ori* activity *in vivo* has also been very recently reached for bacteriophage G4, where a strong temperature-dependent impairment of replication was found after introducing by site-directed mutagenesis point mutations that destabilize intrastrand base-pairing in the *ori* sequence (Lambert *et al.*, 1987). In the latter case, hairpin formation concerns single-stranded and not double-stranded DNA. The possibility should be left open, therefore, that hairpin formation in the petite genomes discussed here may be affected in the single-stranded DNA made during DNA polymerase action. This would not change, however, the basic conclusion that temperature differentially and reversibly affects secondary DNA structures as present during the replication of the mitochondrial genomes of petites 14 and 26.

From a general viewpoint, these results indicate the existence of a novel type of environment–genome interaction, in which reversible changes in higher-order DNA structures are induced with profound consequences on a basic genome function, such as replication. These changes concern genome transconformations which, although non-inheritable, can be maintained for many generations in the presence of the appropriate environmental condition. Genome transconformations can provide, therefore, strong selective advantages or disadvantages and play an important role in evolution, independently of classical mutations, which involve changes in the primary structure of DNA.

It should also be noted that genome transconformations of the type just described are likely to be found in organelle genomes from other poikilothermic organisms and also in prokaryotic genomes, which have a similar nucleoprotein organization and can replicate at widely different temperatures. Moreover, similar phenomena might (1) be induced by other environmental factors; (2) affect other genome functions (e.g. transcription); and also (3) be operative in other organisms.

REFERENCES

Baldacci, G., Cherif-Zahar, B. and Bernardi, G. (1984) The initiation of DNA replication in the mitochondrial genome of yeast. *EMBO J.* **3**, 2115–20.

Bernardi, G. (1982) Evolutionary origin and the biological function of non-coding sequences in the mitochondrial genome of yeast. In: Slonimski, P.P., Borst, P. and Attardi, G. (Eds.) *Mitochondrial Genes*, Cold Spring Harbor Laboratory, Cold Spring Harbor, New York, pp. 269–78.

Bernardi, G. (1983) Genome instability and the selfish DNA issue. *Folia Biol.* **29**, 82–92.

Bernardi, G. and Bernardi, G. (1986) Compositional constraints and genome evolution. *J. Mol. Biol.* **24**, 1–11.

Colin, Y., Baldacci, G. and Bernardi, G. (1985) A new putative gene in the mitochondrial genome of Saccharomyces cerevisiae. *Gene*, **36**, 1–13.

de Zamaroczy, M. and Bernardi, G. (1985) Sequence organization of the mitochondrial genome of yeast—a review. *Gene* **37**, 1–17.

de Zamaroczy, M. and Bernardi, G. (1986a) The GC clusters of the mitochondrial genome of yeast and their evolutionary origin. *Gene* **41**, 1–22.

de Zamaroczy, M. and Bernardi, G. (1986b) The primary structure of the mitochondrial genome of *Saccharomyces cerevisiae*—a review. *Gene* **47**, 155–77.

de Zamaroczy, M. and Bernardi, G. (1987) The AT spacers and the *var*1 genes from the mitochondrial genomes of *Saccharomyces cerevisiae* and *Torulopsis glabrata* : evolutionary origin and mechanism of formation. *Gene* **54**, 1–22.

de Zamaroczy, M., Faugeon-Fonty, G. and Bernardi, G. (1983) Excision sequences in the mitochondrial genome of yeast. *Gene* **21**, 193–202.

de Zamaroczy, M., Faugeron-Fonty, G., Baldacci, G., Goursot, R. and Bernardi, G. (1984) The *ori* sequences of the mitochondrial genome of a wild-type yeast strain : number, location, orientation and structure. *Gene* **32**, 439–57.

Gaillard, C., Strauss, F. and Bernardi, G. (1980) Excision sequences in the mitochondrial genome of yeast. *Nature* **283**, 218–20.

Gennaro, M.L. and Novick, R.R. (1986) Cmp, a *cis*-acting plasmid locus that increases interaction between replication origin and initiator protein. *J. Bacteriol.* **168**, 160–6.

Goursot, R., Goursot, R. and Bernardi, G. (1988) Temperature can reversibly modify the structure and the functional efficiency of *ori* sequences of the mitochondrial genome from yeast. *Gene*, **69**, 141–5.

Lambert, P.F., Kawashima, E. and Reznikoff, W.S. (1987) Secondary structure at the bacteriophage G4 origin of complementary strand DNA synthesis : *in vivo* requirements. *Gene* **53**, 257–64.

Rayko, E., Goursot, R., Cherif-Zahar, B., Melis, R. and Bernardi, G. (1988) Regions flanking *ori* sequènces affect the replication efficiency of the mitochondrial genome of *ori*+ petite mutants from yeast. *Gene* **63**, 213–26.

Rickwood, D., Chambers, J.A.A. and Barat, M. (1981) Isolation and preliminary characterization of DNA–protein complexes from the mitochondria of *Saccharomyces cerevisiae*. *Exp. Cell. Res.* **133**, 1–13.

6 Population Consequences of Genetic Design in Sexually Reproducing Organisms

FREDDY BUGGE CHRISTIANSEN
Department of Ecology and Genetics, University of Aarhus, Aarhus, Denmark

The source of evolutionary change is heritable variation. This variation emerges as essentially random mutations in the genes recording the characteristics of organisms. In the Darwinian theory of evolution a basic assumption is that variation in a population is produced independently of natural selection, and so mutations are unrelated to the appearance, the phenotype, of the organism. This separation of the source of change, mutations, from the evolutionary force is intrinsic to Mendelian inheritance, in that indirect inheritance separates the hereditary material, the genes, from the phenotype of the organism. The neo-Darwinian theory of evolution, which is Darwin's theory including Mendelian inheritance, is to a large extent stretched over this framework, and it constitutes our main conception of the genetic basis of evolutionary change.

Exchange of genetic material among individuals within a species occurs by sexual processes, in higher organisms by sexual reproduction. Exchange of genetic material between organisms of different species rarely occurs in nature, but it may be possible through hybridization between closely related species with sexual reproduction. This kind of genetic exchange is unlikely for more distantly related species.

Genetic entities may, however, be infectiously transmitted from other organisms, a process known as horizontal transfer. This transfer represents a qualitative change in the rigid description of mutational change given above, in that genetic entities that are transmitted infectiously, with high probability, originate from species that survive and reproduce in the environment. The occurrence of horizontal transfer of genetic material may often be described formally as mutualistic symbiosis. For instance, the relationship between bacteria and plasmids conveying resistance to an environmental stress, such as the application of antibiotics, may be described as an interaction between a host and a parasite. The plasmids usually convey

Introduction of Genetically Modified Organisms into the Environment. Edited by Harold A. Mooney and Giorgio Bernardi

a cost to the host bacterium, so without the environmental stress bacterial cells which accidentally lose the plasmid are at an advantage, and the plasmid may be lost from the population. The incorporation and genomic transmission of pieces of foreign genetic material is in the classical genetic sense a mutation, and because the mutation occurs by horizontal transfer and transformation it depends on the transmissible DNA in the environment. Therefore, such a mutation may potentially be directed by the special demands posed by the environment.

Most of these observations are old and well established, but they have to a large extent been set aside as curiosities of procaryotic, or maybe unicellular, organisms. However, the discovery in higher organisms of transmitted DNA sequences that are clearly insertions of foreign origin takes the phenomenon of transformation out of the evolutionary curiosity shop. Nevertheless, it remains to be seen how important this phenomenon has been in the evolution of higher organisms and how frequently the nuclei of germ-line cells in natural populations are exposed to foreign DNA.

Mutation by transformation retains the property that the process of mutation is independent of natural selection; only the range of possible mutations may be influenced by the environment. The main property of this extension of the classical mutation concept is that quantum jumps in the range of capabilities of the organism are allowed; that is, new functions are not necessarily simple modifications of existing functions. Large changes in morphology or physiology are unlikely to be favoured by natural selection, so these are not the interesting jumps. Rather, the interesting transformational mutations are those that open new alleys in the further evolutionary development of the organism.

Genetic variation created by mutations is the source of evolutionary change in nature, and the same genetic variation has been the source for anthropogenic changes of domestic plants and animals by artificial selection. In this way the process of evolution and the process of plant and animal breeding is inherently the same as reflected in Darwin's original formulation of the theory of evolution by natural selection. A breeder chooses among the various evolutionary possibilities open to the particular organism when directed changes are sought, and the choices may produce processes that quantitatively differ from processes in nature. For instance, a plant breeder frequently employs polyploidizations of species or hybridizations of closely related species to change the available hereditary variation, and a contemporary breeder has techniques to facilitate formations of these changes. Many old domestic plants originate from these processes and many naturally occurring species show signs of the occurrence of these processes. Similarly, novel genetic variation may be found at a higher rate if the mutation rate is increased by chemical or radiative treatments, but again this is just a quantitative change of the naturally occurring process. The variation created by the breeder by these methods remains essentially random with respect to the goal of the breeder, as it is expressed in the artificial selection applied on the variation.

The possibility of directed transformation by the exposure of an organism to specific genetic material is a radical change compared to the situation in classical plant and animal breeding. The change is from quantitative management of the processes governing the creation of genetic variation to management of the quality of the genetic variation. A gene identified to have a certain function can be injected into individuals of a given species in a range of known or constructed allelic variants, and the performance of these different variants can be compared in order to form the basis for improvement of the species in the above sense. Therefore, the creation of variation in a breed can be made highly specific, so the production of variation and the artificial selection can both be oriented toward the goal of the breeder (Smith *et al.*, 1987). Essentially the breeder can pick the variant that has the most desired property and use this for the production of an improved breed.

This breeding procedure is perceived to have two main advantages, both contributing to the excitement of the recombinant-DNA methodologies. First, the breadth of variation can be greatly extended over that of traditional breeding procedures. The available genetic variations transgress not only the variation present in the species but also the extended variation that might be mobilized from related species where hybridization is possible. The function of interest may, in principle, be conveyed by genes from any organism having this function. Second, the amount of breeding work needed to isolate the desirable variant in the breed is greatly reduced, because the variant classes are genetically discrete. The traditional breeding procedures utilize an elaborate crossing procedure to recover the desired properties of the original breed in an individual with the desired variant trait.

The first benefit should be gauged against the loss of variation in a species that accompanies any change brought about by breeding or by evolution by natural selection. The second benefit is too optimistic (Franklin, 1987), and it should be recognized as a recombinant-DNA breeding scenario which is basically the scenario of directed transformation in procaryotes. It focuses on the gene as the functional unit and assumes a direct connection between the presence of the gene and the expression of the desired quality. In diploid sexually reproducing species, of course, this has to be modified to take dominance relations into account, but this is a minor consideration because the scenario ideally refers to a true-breeding transformed variety. In self-fertilizing plants a couple of generations of selfing will produce such a variety, and the scenario is in general expected to work. However, the hidden assumption is that the transformation occurs in a functional transmissible genome, which in effect is assuming haploid individuals. In highly inbred species this may be a reasonable assumption, but in outbreeding plants and in animals the situation is not expected to be quite as simple.

6.1 A NEW MUTANT

A mutation is a rare event and for practical purposes each mutation is a unique event. Thus, a mutation of a gene in any organism occurs in a given cell with a given genome. Vegetative proliferation of this mutant cell is not only a proliferation of the mutant gene but also of the genome in which the mutation occurred. Therefore, in species that reproduce without sex and recombination, the proliferation of a given mutant results in loss of the preexisting variation in the population. For a mutation produced by recombinant-DNA methodology this may be overcome by transforming a large number of individuals, but in sexually reproducing organisms the uniqueness is a trap and the loss of variation inevitable. For simplicity let us restrict attention to diploid sexually reproducing organisms.

A mutation of a gene in a diploid organism occurs in a given cell in one of the two chromosomes carrying the gene. Only mutations in germ line cells are of interest from a genetic point of view, and we may assume accordingly that mutations only occur at the gamete stage. Then, after a mutation has occurred we expect to see it initially in heterozygote condition in one individual formed by the union of the gamete where the mutation occurred and another 'normal' gamete. We now have two allelic forms of the gene, the old allele a and the mutated allele A, so the genotype of our unique individual is Aa.

No matter how we want to use this new variant, we need to build a homozygous AA stock. In self-compatible plants this may be done by selfing, where a quarter of the progeny is expected to be of type AA. Otherwise, we need more elaborate crossings. For instance, in animals we cross the unique individual Aa to any homozygote aa to get progeny half of which are of type Aa. After multiplication of the heterozygote genotype the desired homozygote type can be obtained from heterozygote by heterozygote matings. This procedure is elementary and seems uncomplicated. However, in every instance the AA individuals produced are highly inbred, so the stock of AA individuals will have lost a lot of the variation present in the original population. In addition, inbreed individuals of a predominantly outbreeding organism are in general weakened with lower fertility, and we have to take into account this phenomenon of inbreeding depression in the construction of breeds homozygote for the new mutant. Inbreeding depression also may serve to illustrate the magnitude of the loss of variation that occurs during the construction of the breed, in that the loss of performance due to inbreeding depression is linearly related to the increase of the inbreeding coefficient of the individual, i.e. to the increase in the probability that a random locus carries alleles that are identical by descent (see, for example, Falconer, 1981).

The amount of inbreeding produced by the proliferation of a new mutation may be reduced by repeated outcrossing of individuals carrying the mutant, i.e. heterozygote Aa individuals, to a population of aa individuals. This procedure would eventually produce a collection of Aa individuals that reflect the ideal situation that the A gene occurs in a population of genomes which represent the original population. However, the approach to this situation depends on the process of recombination, and this is a very slow process for genes linked to the new mutant A.

6.1.1 THE PROCESS OF RECOMBINATION

To evaluate the process of recombination suppose that an Aa individual is crossed to an aa individual and that one Aa individual is chosen among the offspring to repeat the procedure of outcrossing. Let the initial individual of this crossing scheme carry a stretch of chromosome of length R_1 cM (or map units) to the left of the A gene which is transmitted from the original chromosome where the mutation occurred without being interrupted by an event of recombination. Similarly, let R_2 cM to the right of A descend from the original mutant gamete:

$$\begin{array}{c} R_1 \qquad\qquad\qquad R_2 \\ -\ |\!\!-\!\!-\!\!-\!\!-\!\!-\!\!-\!\!-\!\!-\!\!-\!\!-\!\!-\!\!-\!\!-\!\!-\!\!-\!\!-\!\!-\!\!-\!\!-\!\!|\ - \\ \mathrm{A} \end{array} \tag{6.1}$$

Let r_1 and r_2 be the corresponding recombination frequencies, that is $r_i = R_i/100$, $i = 1,2$. Further, let r'_1 be the recombination frequency describing the length of the left conserved piece of chromosome after one generation. The expected value of r'_1 is then

$$E(r'_1) = (1 - r_1)r_1 + r_1\left(\frac{r_1}{2}\right) \tag{6.2}$$

because with probability $1 - r_1$ no recombination occurs in the conserved piece ($r'_1 = r_1$) and with probability r_1 a recombination occurs leaving on average half the piece. Therefore,

$$E(r'_1) = \left(1 - \frac{r_1}{2}\right)r_1 \tag{6.3}$$

and in the long run the length of the conserved piece of chromosome is expected to decrease by a factor of $1 - r_1/2$ every generation, which means a very slow change when the conserved piece becomes small. Similarly, for the right piece of chromosome and for the total length of the conserved piece of chromosome $r = r_1 + r_2$ we get

$$E(r') = E(r'_1 + r'_2) = \left(1 - \frac{r_1}{2}\right)r_1 + \left(1 - \frac{r_2}{2}\right)r_2 \tag{6.4}$$

which if r_1 and r_2 are not too different is approximately given by $E(r') \approx (1 - r/4)r$ and if r_1 or r_2 is close to zero it is approximately given by $E(r') \approx (1 - r/2)r$.

These deterministic calculations provide a crude picture of the stochastic process, but the order of magnitude of the effect turns out to be correct. The maximal (or initial) length of the chromosome piece is r_1, and we will assume that at most one chiasmata forms within this interval (complete interference). Let $F(r)$ be the probability that the length of the left piece of chromosome in a given generation is less than the value r. Then the corresponding probability in the next generation is

$$F'(r) = F(r) + r[1 - F(r)] \tag{6.5}$$

because a piece with length less than r stays shorter than r and a longer piece is changed into a piece less than r by a recombination within $100r$ crossover units from the A gene. The probability density function, f say, describing the length of the left piece of chromosome in a given generation is therefore given in the next generation by

$$f'(r) = (1-r)f(r) + \int_r^{r_1} f(\rho)\,d\rho \tag{6.6}$$

If initially the length of the chromosome piece is r_1, i.e. the initial probability density function, f_θ, puts all of its mass at the point r_1, then the probability density function, f_t, describing the length of the left piece of chromosome after t generations is given by

$$f_t(r) = (1-r)^t f_\theta(r) + t(1-r)^{t-1} \tag{6.7}$$

The mean length of the conserved piece of chromosome is then

$$E(r_1^{(t)}) = \frac{1 - (1-r_1)^{t+1}}{t+1} \tag{6.8}$$

and the variance is

$$\begin{aligned} V(r_1^{(t)}) = {} & \frac{t}{(t+2)(t+1)^2} - \frac{2t}{(t+1)^2}(1-r_1)^{t+1} \\ & + \frac{2}{t+2}(1-r_1)^{t+2} - \frac{1}{(t+1)^2}(1-r_1)^{2(t+1)} \end{aligned} \tag{6.9}$$

The joint probability density of the left and right pieces, again assuming complete interference, can be found in a similar way as a function of the initial length of the chromosome piece to the left and right of the A gene.

Table 6.1 Conserved piece of chromosome after 10 generations for $r_1 = r_2$

Initial piece r	Distribution: mean ± s.d.	Equation (6.4) rate $1 - r/4$
0.02	0.019 ± 0.002	0.019
0.04	0.036 ± 0.007	0.036
0.06	0.052 ± 0.012	0.052
0.08	0.066 ± 0.017	0.065
0.10	0.078 ± 0.023	0.078
0.12	0.090 ± 0.028	0.089
0.14	0.100 ± 0.034	0.098
0.16	0.109 ± 0.040	0.106
0.18	0.117 ± 0.045	0.114
0.20	0.125 ± 0.051	0.120

Tables 6.1 and 6.2 show the values of $E(r^{(10)})$ in the simple case where $r_1 = r_2$ (Table 6.1) and in the case of initial asymmetry (Table 6.2). The mean values are compared to the deterministic prediction given by an iteration of equation (6.4) with the identification $r' = E(r)$. Both tables show that the deterministic prediction gives a fair picture for judging the order of magnitude of the conserved chromosome piece. Therefore, the conclusion from the deterministic evaluation holds, in that the process of recombination is extremely slow in reducing and eventually removing the piece of chromosome associated with the new mutant. Ten generations of recombination only halves an initial piece of 20 cM around the new mutant.

6.1.2 THE EFFECT OF FINITE POPULATION SIZE

The simple outcrossing scheme considered above replaces random chromosome segments flanking the conserved segment from the population of aa individuals. The A gene is placed in a random chromosome from the same

Table 6.2 Conserved piece of chromosome after 10 generations with initial asymmetry between the left and the right piece of chromosome: $r_1 \neq r_2$ and $r = 0.2$

Initial		Mean	Equation (6.4)
r_1	r_2		
0.10	0.10	0.125	0.120
0.12	0.08	0.123	0.118
0.14	0.06	0.119	0.112
0.16	0.04	0.111	0.102
0.18	0.02	0.099	0.088
0.20	0.00	0.083	0.070

population, so the procedure of recombination can restore the population variation around the A gene at the same time as the A gene is multiplied. However, the amount of variation restored by this procedure is very dependent on the number of independent outcrossing schemes used to produce the Aa individuals that founded the AA stock.

Within the conserved piece of the original chromosome containing the A gene every locus contains a copy of the allele carried in that chromosome. Any locus to the left of A which recombines with the A locus with probability r remain after t generations of outcrossing within the conserved piece with probability

$$\int_{r}^{r_1} f_t(\rho)\, d\rho = (1-r)^t \tag{6.10}$$

when $r \leq r_1$, where r_1 as before is the initial length of the piece. Therefore, if we produce N independent Aa individuals by t generations of outcrossing, then we expect to get

$$N_t(r) = (1-r)^t + [1 - (1-r)^t]N \tag{6.11}$$

independently derived copies of a gene that recombines with the A gene with probability r. Table 6.3 shows this effective number of genes produced from 100 lines of outcrossing in five and ten generations. For five generations of outcrossing the reduction is appreciable for loci within a stretch of about 10 cM around the A gene and after another five generations this region is about halved. Thus, even with very elaborate outcrossing schemes significant loss of variation is expected in the region surrounding the new mutant.

6.1.3 THE INBREEDING DEPRESSION

The loss of natural variation in the region surrounding the new mutant is a

Table 6.3 Effective number of genes produced from 100 outcrossing lines

Recombination frequency r	Generation number	
	$t = 5$	$t = 10$
0	1	1
0.01	6	10
0.02	11	19
0.04	19	34
0.08	35	57
0.16	59	83
0.32	86	98

reflection of the inbreeding at the loci in that region due to the formation of the desired, but inbred, homozygote AA. This inbreeding is expected to produce inbreeding depression in predominantly outbreeding species, and this predicted effect will be compared to data from natural and experimental populations in the next section. The inbreeding effect is linked to the A gene, so the genotypes aa and Aa will not be expected to show any effects, whereas the AA genotype is expected to have a decreased fitness due to the inbreeding depression. Thus, even after elaborate outcrossing the new mutant is expected to act like a recessive deleterious allele unless it has fitness effects of its own. The new mutant acts as a genetic marker for the piece of chromosome around it.

6.1.4 GENETIC HITCH-HIKING

The initial association of a new mutant with the genome in which it arises is a special case of the buildup of linkage disequilibrium in finite populations due to stochastic effects paralleling random genetic drift (Sved, 1971a; Weir and Cockerham, 1977). The persistence of an association after extensive outcrossing parallels the hitch-hiking effect of an advantageous gene on linked genes (Maynard Smith and Haig, 1974; Thomson, 1977), a powerful effect that can even carry recessive lethals to high frequencies (Wagener and Cavilli-Sforza, 1975). The considerations are also closely related to the theory of limits to artificial selection where effects of genetic linkage are taken into account (Hill and Robertson, 1966; Robertson, 1970).

6.2 ASSOCIATIVE FITNESS EFFECTS

Inbreeding depression of viability and fecundity due to homozygosity of chromosomes, or parts of chromosomes, is a well known phenomenon extensively studied in natural populations of *Drosophila* (Dobzhansky *et al.*, 1963; Sved, 1971b; for a review see Lewontin, 1974). The effect of inbreeding in genetically marked chromosome pieces is well illustrated in many experimental *Drosophila* populations. In *Drosophila melanogaster* populations, founded from an inbred stock carrying the recessive ebony mutant allele e and from an inbred stock with the ebony wild-type allele +, Frydenberg (1964) observed initially (10–20 generations) a stable polymorphism indicating that the heterozygote e+ had a higher fitness than both homozygotes ee and ++. However, after some time (30–40 generations) the stability of the polymorphism apparently broke down, and eventually the ebony allele seemed to be on the way out of the populations (after 60–70 generations). Frydenberg interpreted these results as an effect of inbreeding and linkage in the sense discussed above. The e allele acted as a marker of a piece of chromosome from the inbred ebony stock, and the + allele marked a chromosome piece from the inbred wild-type stock, so fitness of the homozygote genotypes at the ebony locus were influenced by the inbreeding depression due to these linked pieces of chromosomes. The fitness of the heterozygote e+, on the other hand, would not be influenced by inbreeding depression, so this genotype is expected to be superior in fitness as long as the inbreeding depression is appreciable. Therefore, the initially stable polymorphism was due to associative overdominance at the ebony locus. As the recombination process reduced the length of the associated chromosome pieces the intrinsic fitness effect of the variation at the ebony locus played a larger role, and the e allele finally behaved as a deleterious allele.

Observations of associative overdominance of genetic markers derived from inbred stocks are very common. An especially interesting experiment is described by Barker (1977). He used two alleles w and w^{b1} at the white locus in *Drosophila melanogaster* in populations under conditions where the intrinsic fitness effects were minimized. The resulting associative effects gave an initial period of strong associative overdominance, but during the time following this period different populations showed different associative fitness effects. After 200 generations two populations remained polymorphic at different frequencies, and these polymorphisms lasted and changed slowly over the next 400 generations. These extended associative fitness effects are expected if the uniqueness of advantageous recombinant chromosomes are taken into account. A favoured recombinant chromosome may cause extensive hitch-hiking and buildup of novel associations between the genetic marker alleles and large pieces of chromosome.

In natural populations similar phenomena have been observed. Barker and East (1980) studied the effect of a release of genetically marked *Drosophila buzzatii* into an isolated natural population. The genetical markers were naturally occurring alleles at three enzyme loci obtained by 37 independent isolations from natural populations. The flies were produced in laboratory populations formed by intercrossing of these 37 lines. The subsequent release was successful in markedly changing the gene frequencies of the alleles in the natural population, but after the experimental release stopped the gene frequencies returned quite rapidly to the original values of the undisturbed population. In a later more extensive experiment at a different location (Barker *et al.*, 1989) a locus perturbed in the same way as before was pushed almost to fixation for the released allele (Figure 6.1). The released allele was kept at a constant high frequency for a long time with intensive release, but again after the experimental release ceased the gene frequency returned quite rapidly to the original value of the undisturbed population (even though the released allele came from about double the number of isolations). The most immediate interpretation of these results is that the released allele is an associative deleterious recessive, as would be expected if it was associated with chromosome pieces showing inbreeding

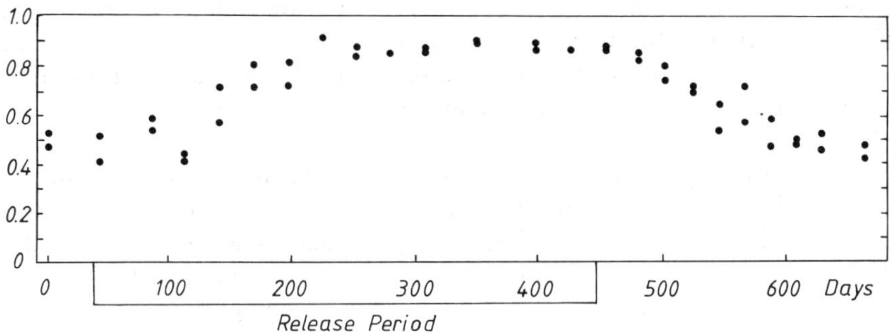

Figure 6.1 The change in allele frequencies at an enzyme locus in a natural population of *Drosophila buzzatii* due to experimental release of flies marked as homozygotes for a naturally occurring allele. The frequency of the allele used as a genetic marker is shown in two sections of the isolated population before (the first two samples), during (the next 13 samples) and after (the last nine samples) the period in which marked flies were released into the population. (Redrawn after Barker *et al.*, 1989.)

depression. This explains the long period with a high gene frequency as due to an immigration-selection balance equilibrium. Thus, either the number of independent isolations of the released allele is too low to average out the associative fitness effect or the proliferation of these flies in the laboratory builds associations between the markers and genes that are advantageous in the laboratory but disadvantageous in nature.

6.3 DISCUSSION

Recombinant-DNA methodology provides the breeder with increased genetic variation in the form of isolated genes. It does not provide an increase in the heritable phenotypic variation which is usually the source of improvement in outbreeding populations, especially of animals. The focus on single genetic elements in the breeding of an animal therefore goes counter to usual animal breeding practice. This may pose the problem that transfer of genes requires knowledge of single gene traits to make it worth the effort (Robertson, 1982), and our knowledge of genes involved in the heritable variation of desirable phenotypic traits is sparse. In addition, genes with a product that seems immediately related to some desirable character may not be the most important genes in the breeding for improvement of that character (Barker, 1985), so the loss of genetic variation in an animal breed following a genetic transformation is a very important cost to be considered in any breeding application of the methodologies of gene transfer. Loss of variation is of course a general concern in any animal breeding practice, and care is usually taken to avoid inbreeding with the concomitant inbreeding depression and general loss of genetic variation. However, the focus on one gene produces inbreeding around that gene even if carefully controlled outbreeding is performed. Further, the focus on one aspect of one trait carries with it the danger of underestimating the general inbreeding effects. Surely, a simplified picture of gene transfer as a quick and easy breeding scenario has to be modified.

The focus on single gene traits in a sense transforms animal breeding into plant breeding, where breeding procedures with a focus on genes are more widespread. In plants the breeding for a particular gene in an outbreeding plant also produces inbreeding effects, but extensive outcrossing and conservation of the original variation during breeding is more practical. Further, the more effective proliferation of plants allows a circumvention of inbreeding problems by producing commercial seeds by a hybridization procedure. However, it is not clear that these procedures in the long run are superior to procedures that are closer to the animal breeding techniques. The concept of uncomplicated and speedy gene transfer will probably be restricted to applications in selfing or in non-sexual plants.

The application of recombinant-DNA methodologies in plant and animal breeding provides exciting possibilities, but the complicated genetic structure of outbreeding populations entails limitations for the technique. However, the rDNA methodologies allow investigations with a focus on effects linked to single genes in relation to characters that show polygenic inheritance. Therefore, the techniques have a great potential as a tool to further the understanding of the genetic basis of these characters. In this way the methodology will be a potential tool for the long-term understanding of the processes in animal and plant breeding and therefore also for our understanding of the process of evolution.

The considerations here are also of relevance in the evaluation of potential applications of recombinant-DNA methodologies in transforming higher organisms in nature, e.g. to make natural enemies of pests more effective. This case is not very different from a usual animal breeding problem. However, one additional concern is appropriate here, namely the fate of a locally released organism with a special discrete genetic difference from the natural population. The experiments with *Drosophila buzzatii* described above show that the modified genes probably will be disadvantageous in nature due to associative fitness effects, so the modified type can be maintained in nature only by the pressure from recurrent immigration due to repeated release. This property is not intrinsic to the modified genes. Suitable recombinants, which are bound to occur in time, may free the modified genes from the breeding burden and permit their ultimate fate to be determined as a function of their intrinsic effects in the natural population.

ACKNOWLEDGEMENTS

The paper benefited from comments on a draft manuscript by Drs J.S.F. Barker and V. Loeschcke. The figure is drawn by Mr A. Jensen.

REFERENCES

Barker, J.S.F. (1977) Population genetics of a sex-linked locus in *Drosophila melanogaster*. I. Linkage disequilibrium and associative overdominance. *Hereditas* **85**, 169–98.

Barker, J.S.F. (1985) Potential contributions of genetics to animal production. In: Proceedings of the 3rd AAAP Animal Science Congress, Korea, pp. 22–32.

Barker, J.S.F. and East, P.D. (1980) Evidence for selection following perturbation of allozyme frequencies in a natural population of *Drosophila. Nature* **284**, 166–8.

Barker, J.S.F., East, P.D. and Christiansen, F.B. (1989) Estimation of migration from a perturbation experiment in natural populations of *Drosophila buzzatii. Biol. J. Linnean Soc.* **37**, 311–34.

Dobzhansky, T., Spassky, B. and Tidwell, T. (1963) Genetics of natural populations. XXXII. Inbreeding and the mutational and balanced loads in natural populations of *Drosophila pseudoobscura*. *Genetics* **48**, 361–73.

Falconer, D.S. (1981) *Introduction to Quantitative Genetics,* 2nd ed., Longman, London.

Franklin, I. (1987) Gene transfer for the Australian sheep industry. In: McGuirk, B.J. (Ed.) *Merino Improvement Program in Australia*, Australian Wool Corporation, Melbourne, pp. 495–509.

Frydenberg, O. (1964) Long-term instability of an *ebony* polymorphism in artificial populations of *Drosophila melanogaster*. *Hereditas* **51**, 198–206.

Hill, W.G. and Robertson A. (1966) The effect of linkage on limits to artificial selection. *Genet. Res.* (*Cambridge*) **8**, 269–94.

Lewontin, R.C. (1974) *The Genetic Basis of Evolutionary Change*, Columbia University Press, New York.

Maynard Smith, J. and Haig, J. (1974) The hitch-hiking effect of a favourable gene. *Genet. Res.*(*Cambridge*) **23**, 23–35.

Robertson, A. (1970) A theory of limits in artificial selection with many linked loci. In: Kojima, K. (Ed.) *Mathematical Topics in Population Genetics*, Springer Verlag, New York, pp. 246–88.

Robertson, A. (1982) Genetic engineering in animal improvement. *Proc. 2nd World Congr. on Genetics Applied to Livestock Production* **6**, 139–45.

Smith, C., Meuwissen, T.H.E. and Gibson, J.P. (1987) On the use of transgenes in livestock improvement. *Animal Breeding Abstracts* **55**, 1–10.

Sved, J. (1971a) Linkage disequilibrium and homozygosity of chromosome segments in finite populations. *Theoret. Population Biol.* **2**, 125–41.

Sved, J. (1971b) An estimate of heterosis in *Drosophila melanogaster*. *Genet. Res.* (*Cambridge*) **18**, 97–105.

Thomson, G. (1977) The effect of a selected locus on linked neutral loci. *Genetics* **85**, 753–88.

Wagener, D.K. and Cavalli-Sforza, L.L. (1975) Ethnic variation in genetic disease: possible roles of hitchhiking and epistasis. *Am. J. Human Genet.* **27**, 348–64.

Weir, B.S. and Cockerham, C.C. (1974) Behavior of pairs of loci in finite monoecious populations. *Theoret. Population Biol.* **6**, 323–54.

7 Evolutionary Consequences of Modifying Cultivated Plants

A. CAPLAN and M. VAN MONTAGU
Institute of Genetics, University of Gent, Gent, Belgium

Within the next few years, a number of new varieties of plants will be tested. Outwardly, these plants will appear identical to the agricultural varieties currently in use. However, whereas the older kinds of plants are sensitive to herbicides, or retarded by virus infection, or ravaged by insect plagues, the new forms will grow more or less unaffected. These new varieties will be protected by one or two genes derived from prokaryotes or viruses and introduced by means of modern techniques of plant molecular biology. This review will (1) summarize how these techniques are applied, (2) describe several examples of the kind of genes being introduced into crop plants, and (3) evaluate the impact the resulting plants might have on agricultural and non-agricultural plant communities.

7.1 GENE TRANSFER BETWEEN BACTERIA AND PLANTS

There are a number of neoplastic growths of plants that are induced by plant-associated bacteria. In 1943, Braun showed that one such type of tumor, called crown galls, continued to grow *in vitro* even when bacteria were no longer present (Braun, 1943). The explanation proved to be that the disease-causing organism, *Agrobacterium tumefaciens*, had somehow transferred to the infected plant cells a new set of genes that promoted dedifferentiation and proliferation (Chilton *et al.*, 1977; Schell *et al.*, 1979). A genetic analysis carried out over the last seven years has revealed the sequence of events leading to this transformation process.

One of the first steps leading to the formation of crown gall tumors is carried out by products of a number of genes lying in the bacterial chromosome. At least three of these cistrons, *chvA*, *chvB*, and *pcsA*, are involved in attaching the bacterium to plant cells (Douglas *et al.*, 1985; Thomashow *et al.*, 1987). During this time, a second set of genes encoding at least 10 proteins comes into play (Stachel and Nester, 1986; Engström

Introduction of Genetically Modified Organisms into the Environment. Edited by Harold A. Mooney and Giorgio Bernardi

et al., 1987). These virulence genes lie within a 35-kilobase region of a very large, tumor-inducing (Ti) plasmid. Only two of them, *virA* and *virG*, are expressed initially. However, the other virulence proteins are induced when the bacterium senses the presence of phenolic compounds, such as acetosyringone (Stachel *et al.*, 1985), that are excreted from wounded plant cells during the course of cell wall repair. The precise functions of all of these proteins remains unresolved, but it is clear that they are responsible for producing a single-stranded DNA copy of the actual tumor-inducing genes and for transferring this copy from bacteria into plant cells (Stachel *et al.*, 1986; Wang *et al.*, 1987) in a process similar to some forms of bacterial conjugation (Buchanan-Wollaston *et al.*, 1987).

The tumor-inducing genes of *Agrobacterium* lie within the T-DNA region of the Ti plasmid. Two of the genes cooperate to produce the plant growth factor, auxin (Inzé *et al.*, 1984; Schröder *et al.*, 1984; Thomashow *et al.*, 1984), and one additional gene is responsible for the synthesis of a second factor, cytokinin (Akiyoshi *et al.*, 1984; Barry *et al.*, 1984). It is the unregulated production of these two hormones that promotes the proliferation of the infected plant cell so that it forms an undifferentiated crown gall tumor. Plant cells, particularly from dicotyledonous plants, have enormous developmental potential. Thus, if either of the auxin-producing genes is deleted, then the crown gall tissue contains an excess of cytokinin relative to auxin with the result that the tumor produces numerous shoots and leaves (Garfinkel *et al.*, 1981; Joos *et al.*, 1983). On the other hand, if the single cytokinin-synthesizing gene is removed, the tumors produce an excess of auxin and some of the tissues differentiate into roots. If all of the T-DNA genes are removed, leaving only the DNA sequences at the very ends of the T-DNA, then no tumors arise (Leemans *et al.*, 1982; Joos *et al.*, 1983; Zambryski *et al.*, 1983). Nevertheless, the remaining parts of the T-DNA (and any new sequence inserted between the ends) are copied, transferred to the plant, and integrated in the nuclear genome in a colinear unit, identical to the copy originally in the bacterium.

The T-DNA of *Agrobacterium* is unique as a gene transfer vector. The integrated form does not undergo successive cycles of replication or integration like a virus or transposon. Moreover, the proteins required for intercellular transfer remain (as far as is known) outside of the plant cell (Otten *et al.*, 1984). As a result, integrated T-DNAs do not increase in number or change position even when cells are reinfected by *Agrobacterium*. In fact, integrated T-DNAs lack intact copies of the DNA sequences that are required by the bacterial virulence genes to process the T-DNA for the DNA transfer operation (Zambryski *et al.*, 1980; Wang *et al.*, 1984). Consequently, once a T-DNA is inserted in the plant genome, it is as stable as other plant sequences (Müller *et al.*, 1987). These properties have contributed to the widespread use of the T-DNA in much of the genetic engineering currently being done with plants.

7.2 THE BASIC PRINCIPLES OF GENETIC ENGINEERING

The genetic improvement of an organism depends on the breeder's ability to manipulate the genetic diversity within the species. Every cross that introduces a desired trait potentially introduces undesirable ones. These latter genes must be removed by subsequent crosses, sometimes over the course of many years.

The alternative approach is to use techniques developed for DNA cloning to first isolate a gene from one plant and then to introduce it into another. In fact, there is no need for the gene to come from the same species. It is possible to take any gene, prokaryotic or eukaryotic, provide it with suitable transcription and translation signals, and introduce it into an organism of another kingdom. In the case of plants, this entails inserting the isolated and/or modified gene between the terminal sequences of a T-DNA and then bringing this into specially modified agrobacteria that no longer contain any other T-DNA sequences (Deblaere *et al.*, 1985; Koncz and Schell, 1986). Using appropriate culture conditions, these bacteria can infect and transfer the artificial T-DNA either to pieces of leaf tissue (Horsch *et al.*, 1985) or to leaf protoplasts (De Block *et al.*, 1984). If the tissues or cells are given suitable amounts of auxin and cytokinin, then they will grow *in vitro* as callus or shoot tissues, and later become fully developed plants when the hormones are removed (Nagata and Takebe, 1970). Three examples of the kinds of genes that can be introduced are presented in the following pages.

7.2.1 CASE 1. PROTECTING PLANTS AGAINST *LEPIDOPTERA* LARVAE

Since the latter half of the 1950s, several companies have been marketing a biological insecticide based on the crystal protein of *Bacillus thuringiensis*. This protein seems to have no effect on vertebrates or most insects, but kills *Lepidoptera* larvae within 3–5 days of ingestion (Carlton and González, 1986). Despite its specificity and effectiveness, it is not extensively used. This is in part due to the fact that it is perceived to have a low degree of field persistence: in order to be fully effective, it is necessary to apply it uniformly, and sometimes repeatedly, over the insect-infected field.

One possible solution to this problem has recently been found. The crystal protein gene was cloned and subsequently modified so that it could be expressed efficiently in plant cells. This modified gene was then introduced, using the technique described earlier, into cells of tobacco (Vaeck *et al.*, 1987) and tomato (Fischhoff *et al.*, 1987), and transgenic plants were regenerated. In order to determine whether the new gene conferred the expected trait, pieces of these plants were fed to tobacco hornworm (*Manduca sexta*), tobacco budworm (*Heliothus virescens*), or corn earworm (*Heliothus zea*) larvae. Most of the tobacco hornworm larvae ceased feeding within one day and died within six days. Fewer of the other two insects died, although the growth of the survivors of all three species was severely retarded.

This sort of approach to biological control has several clear advantages over the use of broadcast insecticides. First, the product can be distributed uniformly through the plant by using transcriptional controls that are equally active ('constitutive') in all tissues. Alternatively, it is possible to produce the protein only in tissues most frequently attacked by caterpillars by linking the gene to tissue-specific regulatory signals. Second, the pesticide is contained within the tissue itself and so there is no risk of inadvertant contamination of uncultivated fields. Third, the protein appears to be highly specific and consequently unlikely to affect directly any insect other than the one for which it is intended.

7.2.2 CASE 2. PROTECTING PLANTS AGAINST VIRUS

Liberal application of insecticides can generally halt, and very often eradicate, insect plagues. The spread of virus infection is less readily controlled. For this reason, considerable effort has gone into isolating crop genotypes that either decrease the chances that the plant will be attacked by the insect or fungus agent that transmits the virus or in some way alter the physiological properties of the plants so as to minimize the spread of the virus after infection (Keen, 1986). Unfortunately, the number of genes known to confer some degree of disease resistance is still lower than the number of diseases that afflict the crops. At present, it is considerably easier to isolate and characterize the disease-causing organism than to search the thousands of varieties of a particular host for the loci conferring immunity.

It is for this reason that a number of strategies have been devised to subvert a virus or a portion of that virus to protecting a plant against infection. One approach has been based on the observation that certain non-pathogenic 'satellite' viruses seem to attenuate the effects of pathogenic forms. The satellite virus may encode its own coat protein (Reichmann, 1964) or be packaged with the products of the pathogenic virus (Schneider, 1971). In either case, once the pathogen has acquired a satellite, the two forms are cotransmitted. The satellite virus cannot replicate without the gene products of an autonomous form. As a result, the viruses reach an equilibrium determined by the efficiency with which the satellite sequesters essential products of the autonomous virus, yet still leaves enough of that product for the pathogen to persist.

This balance can be upset to favour the satellite virus. One way to demonstrate this is to establish the satellite in the plant before introducing a pathogen. Since most plant viruses are composed of RNA, this is done by first producing a DNA copy of the satellite and then providing it with the necessary promoters and transcription termination signals. Modified forms of the satellite of cucumber mosaic virus (CMV) and of tobacco ringspot virus (TobRV) were prepared in this way and put into tobacco (Gerlach *et al.*, 1987; Harrison *et al.*, 1987). These viral genomes were phenotypically silent until plants were infected with the appropriate pathogen. The pathogenic virus then provided the replication functions needed to permit the satellite to multiply. The satellite, in turn, prevented the pathogen from replicating, and greatly reduced the severity of the infection.

This approach to protecting plants against virus infection appears quite promising. The system is self-regulating: the more the pathogen replicates initially, the more the satellite replicates subsequently to halt the spread of the disease. Although every virulent form is not parasitized by a satellite virus, it may be possible to create or select ones in the laboratory by modifying the pathogen itself. Alternatively, some viruses can be controlled by overexpressing negative regulators of viral genes or proteins such as the virus coat protein that can block virus replication (Abel *et al.*, 1986; Loesch-Fries *et al.*, 1987).

7.2.3 CASE 3. INTRODUCING HERBICIDE RESISTANCE INTO CROP PLANTS

The more similar a particular agricultural pest is to the plant under cultivation, the more difficult it is to eradicate. Monocotyledonous weeds, for example, can be killed by certain herbicides, and dicotyledonous weeds by others (Gressel, 1986). Few herbicides, however, are so selective that they can be used to kill monocotyledonous weeds in fields of corn or rice, or dicotyledonous weeds growing amidst tomatoes or soybeans.

One way to circumvent this problem has been to select for particular crop lines that are resistant to commercially important herbicides. This strategy, unfortunately, has not always proved effective, in part because it has depended on preexisting sources of genetic variation. Recombinant DNA technology can be used to broaden this base by bringing together genes from wholly unrelated organisms. Several different approaches are being evaluated currently.

One means of conferring herbicide resistance has been to introduce into a crop bacterial genes that are less sensitive to the effects of the herbicide than the endogenous plant gene. The herbicide glyphosate, for example, blocks the synthesis of aromatic amino acids (Jaworski, 1972). The gene target for this chemical has been cloned from *Escherichia coli* and modified so that it is expressed in tomato (Fillatti *et al.*, 1987) and petunia (Della-Cioppa *et al.*, 1987). The resulting transgenic plants clearly tolerate levels of glyphosate that kill unmodified lines, although the levels of resistance are not yet high enough to be commercially useful.

An alternative type of resistance can be created by using enzymes that inactivate the herbicide before it reaches toxic levels in the cell. For example, it has been possible to isolate a gene from the prokaryote, *Streptomyces hygroscopicus*, that detoxifies the broad spectrum herbicide, phosphinothricin (Murakami *et al.*, 1986). When joined to suitable transcriptional signals, the gene produces enough enzyme to inactivate as much as eight times the amount of herbicide commonly used on agricultural land (De Block *et al.*, 1987). With genes such as this in a plant, it would be possible to use the herbicide phosphinothricin to kill virtually any weed, regardless of its relatedness to the crop species.

7.3 EVALUATING THE IMPACT OF TRANSGENIC PLANTS ON AGRICULTURAL AND NON-AGRICULTURAL ECOSYSTEMS

Although each of the three cases that have been discussed involved the use of recombinant DNA, it is possible that alternative solutions could have been found using conventional plant breeding strategies. Genes conferring resistance to plant hoppers and leaf hoppers have been identified in rice (Khush and Virmani, 1985) and to beetles in summer squash (Dhillon and Sharma, 1987). In a similar way, sets of genes have been identified that confer resistance to specific viral pathogens of cotton and barley (Catherall *et al.*, 1970; Siddig, 1970), while herbicide resistance has been extensively investigated in tobacco (Chaleff and Ray, 1984). Recombinant-DNA techniques simply provide a more direct means to add to existing or potential genetic resources.

There are certain aspects of recombinant-DNA technology that set this art apart from traditional genetic breeding practices. Insect or virus resistance genes identified by conventional breeding programmes have, at best, been transferred only between evolutionarily related species. By contrast, once a successful resistance gene has been obtained by means of recombinant DNA, it can be introduced into a number of very different plant genera. Ironically, the ease with which it is possible to repeat a successful gene transfer may help maintain the genetic diversity within a crop species, whereas cross-breeding tends to homogenize species and reduce variation. If, for example, an indigenous variety of potato was being displaced from its traditional region of cultivation because it was more sensitive than a commercial variety to a particular virus, then it might be possible to cure the deficiency and preserve the genotype by introducing a single new gene. This could not be easily done using genetic crosses without concomitantly disrupting and diluting many of the desirable traits of the indigenous variety.

A second distinguishing feature of recombinant-DNA technology is the sheer number of gene combinations that can be tested in a comparatively short time. For example, one of the most frequently mentioned objectives of genetic modification is the improvement of the nutritional quality of plants. Plant breeders cannot select directly for characters such as protein quality. Instead, they must assay the protein content of hundreds of varieties of a particular grain or tuber, and then, over the course of several generations, combine and accentuate the best traits by crosses between different lines. By comparison, the desired sort of genetic variation in the form of the cloned storage proteins of maize, peas, and potatoes can be introduced in one step into a plant to compensate for amino acid deficiencies. Some of these additions will fail if the new proteins are not produced in sufficient quantities to improve the quality of the crop, or if producing the protein reduces the yields, or if the proteins prove unpalatable to the people they are intended to help. Still, so many different combinations can be tested that some will be found to prove successful. In this event, what would be the ecological consequences of producing and releasing these kinds of organisms?

Traits such as improved nutritional quality are unlikely to confer a selective advantage in natural ecosystems. In fact, this type of attribute is more likely to make a plant more attractive to herbivores so that it will be less competitive in non-agricultural plant communities.

Genes conferring herbicide resistance are similarly unlikely to change the fitness of a plant that has entered a natural habitat. These genes are providing proteins that either duplicate steps already present in a metabolic pathway (Della-Cioppa *et al.*, 1987) or modify highly unusual compounds not normally found in plant cells (De Block *et al.*, 1987). In the absence of the herbicide, either type of gene product is likely to be selectively neutral or slightly deleterious according to the metabolic cost of its production.

In comparison to the previous examples, traits such as insect or disease resistance will confer a selective advantage in any natural habitat. This does not automatically imply that the species will flourish. The *Bacillus* toxin only kills caterpillars, not all insects, while the satellite virus only protects against the co-evolved pathogen (and a few closely related forms), not all viruses. This specificity is qualitatively no different from the insect- and virus-resistance plants selected by conventional procedures and already released. The fact that these plants have not become established in natural ecosystems is an indication that their distribution is not limited solely by one form of disease or predation. The transgenic plants derived from the same stock material are unlikely to behave any differently. Clearly, though, as techniques in genetic engineering improve, it will be possible to modify the utilization of abiotic factors in the environment such as water, light, and minerals. Improvements of this sort may enhance the invasiveness of a species and therefore will have to be applied more cautiously.

7.4 THE DISPERSION OF NEW TRAITS

A commonly voiced concern about releasing genetically modified species is that the new traits will be transmitted from the domesticated organism to ones better able to compete in nature, with unpredictable consequences. Virus and bacterially mediated transfer of genes between unrelated eukaryotes is very rare outside the laboratory, so for the moment it must be assumed that the horizontal dispersal of a gene is governed by the frequency of interspecific hybridization. Consequently, each release must be considered individually and take into account the probability a crop will be grown near reproductively compatible wild species. If it is judged that the chance for hybridization is high or that the reproductive advantage of the new trait will allow it to spread beyond the hybrid zone, then it may be necessary to restrict the fertility of the introduced species. This could be done by introducing the recombinant genes only into male-sterile or reproductively constrained lines. Alternatively, it may be possible to genetically engineer genes for male sterility. These could be linked to the agronomically important traits, even before the latter are introduced, in order to reduce undesirable outcrossing.

7.5 PERSPECTIVES

Few, if any, of the anticipated changes wrought in the next few years by plant genetic engineering will be truly novel. By contrast, the genes introduced into the plants in order to accomplish this will be. To this extent, we should anticipate some changes in the rate of evolution of some species. The species most modified, however, will be those that have already been the most studied and changed during their domestication. Few of these past modifications have made these crops more invasive. It is unlikely that many of the new genes will change this, but the fitness of each modified organism should be assessed in a variety of small-scale tests prior to more general release. At the same time, if the engineered traits are successful, then it should be possible to discontinue the use of certain persistant herbicides or non-selective insecticides, and thereby allow some currently disturbed ecosystems to recover. It can be hoped that population and molecular biologists will cooperate to record and interpret both aspects of the release of a modified organism in order to refine our ability to predict the course of evolution of the organism and the other members of its ecosystem.

REFERENCES

Abel, P.P., Nelson, R.S., De, B., Hoffman, N., Rogers, S.G., Fraley, R.T. and

Beachy, R.N. (1986) Delay of disease development in transgenic plants that express the tobacco mosaic virus coat protein gene. *Science* **232**, 738–43.

Akiyoshi, D.E., Klee, H., Amasino, R.M., Nester, E.W. and Gordon, M.P. (1984) T-DNA of *Agrobacterium tumefaciens* encodes an enzyme of cytokinin biosynthesis. *Proc. Natl Acad. Sci. USA* **81**, 5994–8.

Barry, G.F., Rogers, S.G., Fraley, R.T. and Brand, L. (1984) Identification of a cloned cytokinin biosynthetic gene. *Proc. Natl Acad. Sci. USA* **81**, 4776–80.

Braun, A.C. (1943) Studies on tumor inception in crown gall disease. *Am. J. Bot.* **30**, 674–7.

Buchanan-Wollaston, V., Passiatore, J.E. and Cannon, F. (1987) The *mob* and *oriT* mobilization functions of a bacterial plasmid promote its transfer to plants. *Nature (Lond.)* **328**, 172–5.

Carlton, B.C. and González, J.M. Jr (1986) Biocontrol of insects—*Bacillus thuringiensis*. In: Augustine, P.C., Danforth, H.D. and Bakst, M.R. (Eds.) *Biotechnology for Solving Agricultural Problems*, Martinus Nijhoff Publishers, Dordrecht, pp. 253–72.

Catherall, P.J., Jones, A.T. and Hayes, J.D. (1970) Inheritance and effectiveness of genes in barley that condition tolerance to barley yellow dwarf virus. *Ann. Appl. Biol.* **65**, 147.

Chaleff, R.S. and Ray, T.B. (1984) Herbicide-resistant mutants from tobacco cell cultures. *Science* **223**, 1148–51.

Chilton, M.-D., Drummond, M.H., Merlo, D.J., Sciaky, D., Montoya, A.L., Gordon, M.P. and Nester, E.W. (1977) Stable incorporation of plasmid DNA into higher plant cells: the molecular basis of crown gall tumorigenesis. *Cell* **11**, 263–71.

Deblaere, R., Bytebier, B., De Greve, H., Deboeck, F., Schell, J., Van Montagu, M. and Leemans, J. (1985) Efficient octopine Ti plasmid-derived vectors for *Agrobacterium*-mediated gene transfer to plants. *Nucl. Acids Res.* **13**, 4777–88.

De Block, M., Herrera-Estrella, L., Van Montagu, M., Schell, J. and Zambryski, P. (1984) Expression of foreign genes in regenerated plants and their progeny. *EMBO J.* **3**, 1681–9.

De Block, M., Botterman, J., Vandewiele, M., Dockx, J., Thoen, C., Gosselé, V., Movva, R., Thompson, C., Van Montagu, M. and Leemans, J. (1987) Engineering herbicide resistance in plants by expression of a detoxifying enzyme. *EMBO J.* **6**, 2513–18.

Della-Cioppa, G., Bauer. S.C., Taylor, M.L., Rochester, D.E., Klein, B.K., Shah, D.M. Fraley, R.T. and Kishore, G.M. (1987) Targeting a herbicide-resistant enzyme from *Escherichia coli* to chloroplasts of higher plants. *Bio/technology* **5**, 579–84.

Dhillon, N.P.S. and Sharma, B.R. (1987) Genetics of resistance of red pumpkin beetle (*Aulacophora foveicollis*) in summer squash (*Cucurbita pepo* L.). *Theor. Appl. Genet* **5**, 711–15.

Douglas, C.J., Staneloni, R.J., Rubin, R.A. and Nester, E.W. (1985) Identification and genetic analysis of an *Agrobacterium tumefaciens* chromosomal virulence region. *J. Bacteriol.* **161**, 850–60.

Engström, P., Zambryski, P., Van Montagu, M. and Stachel, S. (1987) Characterization of *Agrobacterium tumefaciens* virulence proteins induced by the plant factor acetosyringone. *J. Molec. Biol.* **197**, 635–45.

Fillatti, J.J., Kiser, J., Rose, R., and Comai, L. (1987) Efficient transfer of a glyphosate tolerance gene into tomato using a binary *Agrobacterium tumefaciens* vector. *Bio/technology* **5**, 726–30.

Fischhoff, D.A., Bowdish, K.S., Perlak, F.J., Manone, P.G., McCormick, S.M., Niedermeyer, J.G., Fean, D.A., Kusano-Kretzmer, K., Mayer, E.J., Rochester,

D.E., Rogers, S.G. and Fraley, R.T. (1987) Insect tolerant transgenic tomato plants. *Bio/technology* **5**, 807–13.

Garfinkel, D.J., Simpson, R.B., Ream, L.W., White, F.F., Gordon, M.P. and Nester, E.W. (1981) Genetic analysis of crown gall: fine structure map of the T-DNA by site-directed mutagenesis. *Cell* **27**, 143–53.

Gerlach, W.L., Llewellyn, D. and Haseloff, J. (1987) Construction of a plant disease resistance gene from the satellite RNA of tobacco ringspot virus. *Nature (Lond.)* **328**, 802–5.

Gressel, J. (1986). The challenges of conferring herbicide resistance on crops. In: Augustine, P.C., Danforth, H.D. and Bakst, M.R. (Eds.) *Biotechnology for Solving Agricultural Problems*, Martinus Nijhoff Publishers, Dordrecht, pp. 369–84.

Harrison, B.D., Mayo, M.A. and Baulcombe, D.C. (1987) Virus resistance in transgenic plants that express cucumber mosaic virus satellite RNA. *Nature (Lond.)* **328**, 799–802.

Horsch, R.B., Fry, J.E., Hoffmann, N.L., Eichholtz, D., Rogers, S.G. and Fraley, R.T. (1985) A simple and general method for transferring genes into plants. *Science* **227**, 1229–31.

Inzé, D., Follin, A., Van Lijsebettens, M., Simoens, C., Genetello, C., Van Montagu, M. and Schell, J. (1984) Genetic analysis of the individual T-DNA genes of *Agrobacterium tumefaciens*; further evidence that two genes are involved in indole-3-acetic acid synthesis. *Mol. Gen. Genet.* **194**, 265–74.

Jaworski, E. (1972). The mode of action of *N*-phosphonomethylglycine: inhibition of aromatic amino acid biosynthesis. *J. Agric. Food Chem.* **20**, 1195–8.

Joos, H., Inzé, D., Caplan, A., Sormann, M., Van Montagu, M. and Schell, J. (1983) Genetic analysis of T-DNA transcripts in nopaline crown galls. *Cell* **32**, 1057–67.

Keen, N.T. (1986) Novel approaches to improving disease resistance in plants. In: Augustine, P.C., Danforth, H.D. and Bakst, M.R. (Eds.) *Biotechnology for Solving Agricultural Problems*, Martinus Nijhoff Publishers, Dordrecht, pp. 223–36.

Khush, G.S. and Virmani, S.S. (1985) Breeding rice for disease resistance. In: Russell, G.E. (Ed.) *Plant Breeding*, Vol. 1, Cambridge University Press, Cambridge, pp. 239–80.

Koncz, C. and Schell, J. (1986) The promoter of TL-DNA gene *5* controls the tissue-specific expression of chimaeric genes carried by a novel type of *Agrobacterium* binary vector. *Mol. Gen. Genet.* **204**, 383–96.

Leemans, J., Deblaere, R., Willmitzer, L., De Greve, H., Hernalsteens, J-P., Van Montagu, M. and Schell, J. (1982) Genetic identification of functions of TL-DNA transcripts in octopine crown galls. *EMBO J.* **1**, 147–52.

Loesch-Fries, L.S., Merlo, D., Zinnen, T., Burhop, L., Hill, K., Krahn, K., Jarvis, N., Nelson, S. and Halk, E. (1987) Expression of alfalfa mosaic virus RNA 4 in transgenic plants confers virus resistance. *EMBO J.* **6**, 1845–51.

Müller, A.J., Mendel, R.R., Schiemann, J., Simoens, C. and Inzé, D. (1987) High meiotic stability of a foreign gene introduced into tobacco by *Agrobacterium*-mediated transformation. *Mol. Gen. Genet.* **207**, 171–5.

Murakami, T., Anzai, H., Imai, S., Satoh, A., Nagaoka, K. and Thompson, C.J. (1986) The bialaphos biosynthetic genes of *Streptomyces hygroscopicus*: molecular cloning and characterization of the gene cluster. *Mol. Gen. Genet.* **205**, 42–50.

Nagata, T. and Takebe, I. (1970) Cell wall regeneration and cell division in isolated tobacco mesophyll protoplasts. *Planta* **92**, 301–8.

Otten, L., De Greve, H., Leemans, J., Hain, R., Hooykaas, P.J.J. and Schell, J. (1984) Restoration of virulence of Vir region mutants of *Agrobacterium tumefaciens* strain B6S3 by coinfection with normal and mutant *Agrobacterium* strains. *Mol. Gen. Genet.* **195**, 159–63.

Reichmann, M.E. (1964) The satellite necrosis virus: a single protein and its genetic code. *Proc. Natl Acad. Sci. USA* **52**, 1009–17.

Schell, J., De Beuckeleer, M., De Block, M., De Greve, H., Depicker, A., De Vos, G., De Vos, R., De Wilde, M., Dhaese, P., Dobbelaere, M.R., Engler, G., Genetello, C., Hernalsteens, J-P., Holsters, M., Jacobs, A., Messens, E., Seurinck, J., Silva, A., Van Haute, E., Van Montagu, M., Van Vliet, F., Villarroel, R. and Zaenen, I. (1979) Crown gall: transfer of bacterial DNA to plants via the Ti-plasmid. In: Hall, T.C. and Davies, J. (Eds.) *Nucleic Acids in Plants*, Vol. II, CRC Press, Boca Raton, pp. 195–210.

Schneider, I.R. (1971) Characteristics of a satellite-like virus of tobacco ringspot virus. *Virology* **45**, 108–22.

Schröder, G., Waffenschmidt, S., Weiler, E.W. and Schröder, J. (1984) The T-region of Ti plasmids codes for an enzyme synthesizing indole-3-acetic acid. *Eur. J. Biochem.* **138**, 387–91.

Siddig, M.A. (1970) Breeding for leaf curl resistance in Sakel cotton. In: Siddig, M.A. and Hughes, L.C. (Eds.) *Cotton Growth in the Geziza Environment*, Agricultural Research Corporation, Sudan, p. 153.

Stachel, S.E. and Nester, E.W. (1986) The genetic and transcriptional organization of the *vir* region of the A6 Ti plasmid of *Agrobacterium tumefaciens*. *EMBO J.* **5**, 1445–54.

Stachel, S.E., Messens, E., Van Montagu, M. and Zambryski, P. (1985) Identification of the signal molecules produced by wounded plant cells that activate T-DNA transfer in *Agrobacterium tumefaciens*. *Nature* (*Lond.*) **318**, 624–9.

Stachel, S.E., Timmerman, B. and Zambryski, P. (1986) Generation of single-stranded T-DNA molecules during the initial stages of T-DNA transfer from *Agrobacterium tumefaciens* to plant cells. *Nature* (*Lond.*) **322**, 706–12.

Thomashow, L.S., Reeves, S. and Thomashow, M.F. (1984) Crown gall oncogenesis: evidence that a T-DNA gene from the *Agrobacterium* Ti plasmid pTiA6 encodes an enzyme that catalyzes synthesis of indoleacetic acid. *Proc. Natl Acad. Sci. USA* **81**, 5071–5.

Thomashow, M.F., Karlinsey, J.R. Marks, J.R. and Hurlbert, R.E. (1987) Identification of a new virulence locus in *Agrobacterium tumefaciens* that affects polysaccharide composition and plant cell attachment. *J. Bacteriol.* **169**, 3209–16.

Vaeck, M., Reynaerts, A., Höfte, H., Jansens, S., De Beuckeleer, M., Dean, C., Zabeau, M., Van Montagu, M. and Leemans, J. (1987) Insect resistance in transgenic plants expressing modified *Bacillus thuringiensis* toxin genes. *Nature* (*Lond.*) **328,** 33–7.

Wang, K., Herrera-Estrella, L., Van Montagu, M. and Zambryski, P. (1984) Right 25-bp terminus sequences of the nopaline T-DNA is essential for and determines direction of DNA transfer from *Agrobacterium* to the plant genome. *Cell* **38**, 455–62.

Wang, K., Stachel, S., Timmerman, B., Van Montagu, M. and Zambryski, P. (1987) Site-specific nick in the T-DNA border sequence following *vir* gene expression in *Agrobacterium*. *Science* **235**, 587–91.

Zambryski, P., Holsters, M., Kruger, K., Depicker, A., Schell, J., Van Montagu,

M. and Goodman, H.M. (1980). Tumor DNA structure in plant cells transformed by *A. tumefaciens*. *Science* **209**, 1385–91.

Zambryski, P., Joos, H., Genetello, C., Leemans, J., Van Montagu, M. and Schell, J. (1983) Ti plasmid vector for the introduction of DNA into plant cells without alteration of their normal regeneration capacity. *EMBO J.* **2**, 2143–50.

8 Organization and Dynamics of the *Rhizobium* Genome. A Basis for Introducing Novel Arrangements of Genetic Information into the Environment

RAFAEL PALACIOS, ESPERANZA MARTINEZ, MARGARITA FLORES, DAVID ROMERO, SUSANA BROM, GUILLERMO DÁVILA
Departamento de Genética Molecular, Centro de Investigación sobre Fijación de Nitrógeno, Universidad Nacional Autónoma de México, Cuernavaca, Morelos, Mexico

DANIEL PIÑERO
Departamento de Ecología, Instituto de Biología, Universidad Nacional Autónoma de México, Mexico

8.1 INTRODUCTION

Current progress on the knowledge of the genetic information involved in the symbiotic process will certainly provide in the near future genetically improved *Rhizobium* strains. If these bio-engineered bacteria are to be delivered into natural environments, the fate of these novel introductions should be carefully considered. Concerns have been raised mainly on the transfer of genetic information from introduced strains to pathogenic soil inhabitants or to strains that could eventually become pathogens, and also on the modification of the genetic information subsequent to the release of bacteria in the soil. It is imperative to learn the rules of the behaviour of introduced genetic information in natural environments, from the point of view of its stability and of the frequencies of exchange of information between different organisms.

Symbiotic nitrogen fixation is the main biological entrance of fixed nitrogen in the biosphere. This biological process is only performed by procaryotes,

Introduction of Genetically Modified Organisms into the Environment. Edited by Harold A. Mooney and Giorgio Bernardi

among these *Rhizobium* spp. have attracted special interest. *Rhizobium* strains have been used as inoculants in substitution of nitrogen fertilizer for a long time. In the following pages we will review several aspects of *Rhizobium*–legume symbiosis as well as some characteristics of the general organization and dynamics of the *Rhizobium* genome. We will also present an approach to evaluate the exchange of genetic information among strains in nature.

The bacteria of the genus *Rhizobium* form nitrogen-fixing nodules on the roots of legumes. These symbiotic interactions involve a differentiation process of both partners, the bacteria, and the plant which forms a new organ, the nodule. This process has been extensively studied in several legumes such as alfalfa, clover, pea, soybean, bean, and siratro. The establishment of the symbiosis includes several steps. The bacteria attach to the root hairs or to root cells that will become root hairs; *Rhizobium* cell surface polysaccharides have been considered to mediate this attachment. Rhizobia induce root hair curling and then a tubular structure of plant origin, the infection thread, is gradually formed, leading the bacteria to the cortical cells. There, the infection thread branches off and the bacteria are delivered into the plant cytoplasm surrounded by a membrane of host origin. The bacteria differentiate into bacteroids that fix nitrogen and that interact metabolically with the plant.

During symbiosis, specific sets of both bacteria and plant genes are expressed, establishing the basis for the differentiation process. Recent advances in molecular genetics have greatly contributed to our understanding of the *Rhizobium*–legume interaction. Several symbiotic genes have been identified and isolated; their expression under different conditions has also been studied.

Among others, the early nodulation genes (*nod*), host specificity genes (*hsn*), and nitrogen-fixation genes (*nif*) have been analysed (Long *et al.*, 1982; Kondorosi *et al.*, 1984; Downie *et al.*, 1985; Schofield *et al.*, 1984; Ramakrishnan *et al.*, 1986; Lamb and Hennecke, 1986; Ruvkun and Ausubel, 1980; Hennecke, 1981; Corbin *et al.*, 1982; Torok and Kondorosi, 1981; Ma *et al.*, 1982; Scott *et al.*, 1983, Quinto *et al.*, 1982). In the near future a comprehensive scheme of the bacterial genes that participate in the symbiotic process will certainly emerge.

8.2 A DYNAMIC CONCEPT OF THE *RHIZOBIUM* GENOME

Our current model for the *Rhizobium* genome can be stated as follows. A single cell at a given time has a particular arrangement of genetic information. Two forces participate to increase the diversity of this arrangement of the genome in an ecological time scale: its internal dynamics and the exchange of genetic information with other bacterial strains in nature. The *Rhizobium* genome contains a large amount of repeated DNA sequences that might act as points for recombination resulting in genomic rearrangements. Since these rearrangements are frequently observed under laboratory conditions, a cell gives rise to a population of similar but not necessarily identical organisms. This means that even a single bacterial colony will present genetic heterogeneity. The collection of strains that exchange genetic information among them should be considered as a genomic pool equivalent to the concept of genospecies that Reanney (1978) introduced for bacterial populations.

8.3 REITERATED DNA SEQUENCES IN *RHIZOBIUM*

Repeated DNA sequences are a general characteristic of eukaryotic genomes, but in prokaryotes only in the case of the archaebacteria, *Halobacterium halobium* and *Halobacterium volcanii*, has DNA reiteration been reported as a common genomic feature (Sapienza and Doolittle, 1982).

However, we have detected a large amount of repeated DNA elements in *Rhizobium phaseoli*, the symbiont of the common bean plant *Phaseolus vulgaris*, in *Rhizobium meliloti*, the symbiont of alfalfa, and also in *Agrobacterium tumefaciens*, a plant pathogen taxonomically closely related to *Rhizobium* (Flores *et al.*, 1987). We calculate about two hundred repeated DNA families in the genome of some strains of *R. phaseoli*. The family size was usually small, from two to five elements, but some have more than ten elements. *Rhizobium* and *Agrobacterium* contain large plasmids in addition to the chromosome. The analysis of two *R.* phaseoli strains indicates that DNA reiteration is not confined to a particular replicon but is a property of the whole genome.

The nature of some of the repeated elements has been established in various *Rhizobium* strains. The nitrogenase structural genes are reiterated in *Rhizobium phaseoli* (Quinto *et al.*, 1985), in *Rhizobium fredii*, symbiont of some varieties of soybean (Prakash and Atherly, 1984), in the broad host range strain ANU240 (Morrison *et al.*, 1983), in strain ORS571 that produces stem and root nodules in *Sesbania* (Norel *et al.*, 1985; Donald *et al.*, 1986), and in strains originally isolated from different species of *Phaseolus* and from *Pachyrhizus erosus* (Martinez *et al.*, 1985).

Other examples of repeated DNA sequences in *Rhizobium* and *Bradyrhizobium* (slow-growing rhizobia) strains include insertion sequences (Kaluza *et al.*, 1985; Ruvkun *et al.*, 1982), an early nodulation gene (*nodD*) (Gottfert *et al.*, 1986), and *nif* and *nod* regulatory regions (Better *et al.*, 1983; Rostas *et al.*, 1986).

8.4 GENOMIC REARRANGEMENTS IN RHIZOBIUM

It is a common observation that *Rhizobium* strains can generate variability in regard to colony morphology and symbiotic properties. This has led to certain practices diminishing the possibility of losing some important characteristics. The preservation of strains and their use for inoculation are usually performed without the isolation of single colonies. Strains that are supposed to be homogeneous are commonly subjected to symbiotic cycles in order to select for clones presenting the desired properties.

Futhermore, several reports indicate that when exposed to certain stress conditions or to genetic manipulations, *Rhizobium* can present genomic rearrangements (Berry and Atherly, 1984; Christensen anhd Schubert, 1983; Djordjevic *et al.*, 1982; Kaluza *et al.*, 1985; Soberón-Chávez *et al.*, 1986; Wang *et al.*, 1986; Zurkowski, 1982). We have analysed genomic rearrangements that occur at high frequency under commonly used laboratory conditions that are not considered to cause stress in bacterial populations. The experimental approach was similar to that reported by Sapienza *et al.* (1982) demonstrating the presence of frequent genomic rearrangements in the archaebacteria *H. halobium* and *H. volcanii*. Descendants of a single cell were analysed in regard to the hybridization patterns obtained using as probes recombinant plasmids that reveal repeated DNA sequences. Differences in the hybridization pattern indicate that a genomic rearrangement occurred from the time the culture was a single cell to the time of harvesting individual colonies derived from such a cell. With some of the hybridization probes used, genomic rearrangements were observed at a frequency of 1–5% in 20 generations (Flores *et al.*, 1988).

The mechanisms responsible for the genomic rearrangements observed are not known at the present time. Homologous recombination between different elements of reiterated DNA families could be involved as well as the transposition of insertion sequences. Regardless of the underlying mechanisms and of the actual frequency of genomic rearrangements, our data support the hypothesis that *Rhizobium* strains present internal genomic dynamics that are continually generating subpopulations of similar but not identical organisms.

8.5 COMPARTMENTALIZATION OF THE *RHIZOBIUM GENOME*

The *Rhizobium* genome is partitioned into different compartments, the chromosome, and several plasmids of high molecular weight on the order of 10^2–10^3 kilobase pairs. A usual finding in *Rhizobium* strains of different host specificities is that the nitrogenase structural genes and the nodulation genes are found on a single large plasmid, the symbiotic (*sym*) plasmid

(Hombrecher *et al.*, 1981, Prakash *et al.*, 1981; Rosenberg *et al.*, 1981). Of particular interest is the fact that host-specific determinants are, as well, present in the symbiotic plasmid. In the laboratory, plasmids can be transferred among *Rhizobium* strains with different specificities and it is frequently observed that strains acquire the specificity of the corresponding *sym* plasmid (Hooykaas *et al.*, 1981; Johnston *et al.*, 1978). Moreover, *Rhizobium* plasmids have been transferred to *Agrobacterium tumefaciens* and the recipients have gained the ability to initiate and in some cases establish a symbiotic process with the appropriate host legumes (Hooykaas *et al.*, 1982; Kondorosi *et al.*, 1982; van Brussel *et al.*, 1982; Martínez *et al.*, 1987; Brom *et al.*, 1988). A fundamental question is whether these transfers occur naturally in soil populations (see below).

Besides those carried by the *sym* plasmid, genes relevant to nodulation and nitrogen fixation have been found in the chromosome and in other plasmids. It is not known where genes necessary for other important traits such as saprophytic or host competitive ability reside.

8.6 EXCHANGE OF GENETIC INFORMATION AMONG SOIL BACTERIA

The current *Rhizobium* systematics considers it to be a member of the bacterial family *Rhizobiaceae*. This family is composed of four different genera: *Rhizobium, Bradyrhizobium, Agrobacterium* and *Phylobacterium* (Jordan, 1984). These gram-negative bacteria are able to interact with plants eliciting cortical hypertrophies. *Rhizobium* and *Bradyrhizobium* species form nitrogen-fixing nodules on the roots of many leguminous species, and in some cases on the stems of certain leguminous trees. *Phylobacterium* species produce nodules on the leaves of certain plants. *Agrobacterium* species (with the exception of *Agrobacterium radiobacter*) produce gall hypertrophies on stems and roots of various dicotyledonous plants. Bacteria of the genera *Rhizobium* and *Bradyrhizobium* were originally classified on the basis of host nodulation ability. Criticism to this host-specificity-based taxonomy arose early (Wilson, 1944). Its major limitations have been pinpointed frequently. In particular, these are the existence of cross infection, the low representativity of host plants involved in the classification (around 1% of the described nodulated leguminous species), and, more recently, the discovery in *Rhizobium* species of plasmids carrying symbiotic determinants.

Results from numerical taxonomy (Graham, 1964; Moffett and Colwell, 1968), DNA hybridization (Gibbins and Gregory, 1972; Jarvis *et al.*, 1980; Hollis *et al.*, 1981), and protein patterns (Roberts *et al.*, 1980) of bacteria of the genus *Rhizobium* indicate that different clusters may be defined on the basis of chromosomal characters. The approach of bacterial population genetics has allowed the definition of bacterial clusters based on chromosomal characters and to investigate the level of chromosomal recombination occurring in natural populations (Selander, 1985; Piñero *et al.*, 1988). This approach could be successfully used to understand the fate of strains introduced to the soil.

A critical problem is to define the boundaries of specific plasmids, in particular the symbiotic plasmids, in regard to chromosomal clusters. This problem may be approached by comparing dendrograms based on chromosomal characters with those based on plasmidic characters.

The mismatch of the two types of dendrograms would be an indication of the exchange of plasmids among certain chromosomal clusters, while the dendrograms superposition will indicate the confinement of plasmids. In our current studies dendrograms are being established for both the chromosomes and symbiotic plasmids. These dendrograms have been made by using electrophoretic mobility of enzymes for the chromosome and restriction fragment length polymorphism for plasmids. Preliminary experiments indicate that these methodologies are sensitive enough to detect differences in very closely related bacteria. For a particular type of *Rhizobium phaseoli* we have found evidence for plasmid exchange. These studies will be extended to cover different soil bacteria, both symbiotic and pathogenic, with the aim of defining the genospecies clusters.

8.7 CONCLUDING REMARKS

At the present time, there are studies indicating that plasmid transfer among strains may occur in nature (Schofield *et al.*, 1987; Broughton *et al.*, 1987; Richaume *et al.*, 1989). However, a linkage between plasmids and chromosomes has been observed (Young and Wexler, 1988), indicating that there may be an adaptation of plasmids and chromosomes and that the boundaries for plasmid spread may be somehow restricted. More experiments will be needed to define new guidelines for the safe and successful introduction of strains in agricultural fields. These guidelines should be established based on the knowledge of the ecology of bacterial populations in soil and of the rules operating in ecological time in regard to the stability and interchange of genetic information.

ACKNOWLEDGEMENTS

We acknowledge Lorenzo Segovia for reviewing the manuscript. This work was supported in part by the US National Academy of Sciences/National Research Council by means of a grant from the US Agency for International Development, by a grant C II*-0104-MEX(A) from La Communauté

Économique Européenne and by grants from the Consejo Nacional de Ciencia y Tecnologia (México).

REFERENCES

Berry, J.O. and Atherly, A.G. (1984) Induced plasmid-genome rearrangements in *Rhizobium japonicum*. *J. Bacteriol.* **157**, 218–24.

Better, M., Lewis, B., Corbin, D., Ditta, G. and Helinski, D. (1983) Structural relationships among *Rhizobium meliloti* symbiotic promoters. *Cell*, **35**, 479–85.

Brom, S., Martínez, E., Dávila, G. and Palacios, R. (1988) Narrow and broad host range symbiotic plasmids of *Rhizobium* spp. strains that nodulate *Phaseolus vulgaris*. *Apl. Environ. Microbil.* **54**, 1280–3.

Broughton, W.J., Samrey, U. and Stanley, J. (1987) Ecological genetics of *Rhizobium meliloti*: symbiotic plasmid transfer in the *Medicago sativa* rhizosphere. *FEMS Microbiol. Lett.*, **40**, 251–5.

Christensen, A.H. and Schubert, K.R. (1983) Identification of a *Rhizobium trifolii* plasmid coding for nitrogen fixation and nodulation genes and its interaction with pJB5JI, a *rhizobium leguminosarum* plasmid. *J. Bacteriol.* **156**, 592–9.

Corbin, D., Ditta, G. and Helinski, D.R. (1982) Clustering of nitrogen fixation (*nif*) genes in *Rhizobium meliloti*. *J. Bacteriol.* **149**, 221–118.

Djordjevic, M.A., Zurkowski, W. and Rolfe, B.G. (1982) Plasmids and stability of symbiotic properties of *Rhizobium trifolii*. *J. Bacteriol.* **151**, 560–8.

Donald, R.G.K., Nees, D.W., Raymond, K.C., Loroch, I.A. and Ludwig, R.A. (1986) Characterization of three genomic loci encoding *Rhizobium* sp. strain ORS571 N_2-fixation genes. *J. Bacteriol.* **165**, 72–81.

Downie, J.A., Knight, C.D., Johnston, A.W.B. and Rossen, L. (1985) Identification of genes and gene products involved in the nodulation of peas by *Rhizobium leguminosarum*. *Mol. Gen. Genet.* **198**, 255–62.

Flores, M., González, V., Brom, S., Martínez, E., Piñero, D., Romero, D., Dávila, G. and Palacios, R. (1987) Reiterated DNA sequences in *Rhizobium* and *Agrobacterium* spp. *J. Bacteriol.* **169**, 5782–8.

Flores, M., González, V., Pardo, M.A. Leija, A., Martínez, E., Romero, D., Piñero, D., Dávila, G. and Palacios, R. (1988) Genomic instability in *Rhizobium phaseoli*. *J. Bacteriol.* **170**, 1191–6.

Gibbins, A.M. and Gregory, K.F. (1972) Relatedness among *Rhizobium* and *Agrobacterium* species determined by three methods of nucleic acid hybridization. *J. Bacteriol.* **111**, 129–41.

Gottfert, M., Horvath, B., Kondorosi, E., Putnoky, P., Rodriguez-Quiñones, F. and Kondorosi, A. (1986) At least two *nodD* genes are necessary for efficient nodulation of alfalfa by *Rhizobium meliloti*. *J. Mol. Biol.* **191**, 411–20.

Graham, P.H. (1964) The application of computer techniques to the taxonomy of the root-nodule bacteria of legumes. *J. Gen. Microbiol.* **35**, 511–17.

Hennecke, H. (1981) Recombinant plasmids carrying nitrogen fixation genes from *Rhizobium japonicum*. *Nature* (*Lond.*) **291**, 354–5.

Hollis, A.B., Kloos, W.E. and Elkan, G.H. (1981) DNA:DNA hybridization studies of *Rhizobium japonicum* and related Rhizobiaceae. *J. Gen. Microbiol.* **123**, 215–22.

Hombrecher, G., Brewin, N.J. and Johnston, A.W.B. (1981) Linkage of genes for nitrogenase and nodulation ability on plasmids in *R. leguminosarum* and *R. phaseoli*. *Mol. Gen. Genet.* **182**, 133–6.

Hooykaas, P.J.J., van Brussel, A.A.N., den Dulk-Ras, H., van Slogteren, G.M.S. and Schilperoort, R.A. (1981) Sym plasmid of *Rhizobium trifolii* expressed in different rhizobial species and *Agrobacterium tumefaciens*. *Nature* (*Lond.*) **291**, 351–3.

Jarvis, B.D.W., Dick, A.G. and Greenwood, R.M. (1980) Deoxyribonucleic acid homology among strains of *Rhizobium trifolii* and related species. *Int. J. Syst. Bacteriol.* **30**, 42–52.

Johnston, A.W.B., Beynon, J.L., Buchanan-Wollaston, A.V., Setchell, S.M., Hirsch, P.R. and Beringer, J.E. (1978) High frequency transfer of nodulating ability between strains and species of *Rhizobium*. *Nature* (*Lond.*) **276**, 634–6.

Jordan, D.C. (1984). Rhizobiaceae. In: Kreig, N.R. and Holt, J.C. (Eds.) *Bergey's Manual for Determinative Bacteriology*, 9th ed., Williams and Wilkins, Baltimore, pp. 234–44.

Kaluza, K., Hahn, M. and Hennecke, H. (1985) Repeated sequences similar to insertion elements clustered around the nif region of the *Rhizobium japonicum* genome. *J. Bacteriol.* **162**, 535–42.

Kondorosi, A., Kondorosi, E., Pankhurst, C.E., Broughton, W.J. and Banfalvi, Z. (1982) Mobilization of a *Rhizobium meliloti* megaplasmid carrying nodulation and nitrogen fixation genes into other rhizobia and *Agrobacterium*. *Mol. Gen. Genet.* **188**, 433–9.

Kondorosi, E., Banfalvi, Z. and Kondorosi, A. (1984) Physical and genetic analysis of a symbiotic region of *Rhizobium meliloti*: identification of nodulation genes. *Mol. Gen. Genet.* **193**, 445–52.

Lamb, J.W. and Hennecke, H. (1986) In *Bradyrhizobium japonicum* the common nodulation genes, *nod*ABC, are linked to *nif* A and *Fix A*. *Mol. Gen. Genet.* **202**, 512–17.

Long, S.R., Buikema, W.J. and Ausubel, F.M. (1982) Cloning of *Rhizobium meliloti* nodulation genes by direct complementation of Nod$^-$ -mutants. *Nature* (*Lond.*) **298**, 485–8.

Ma, Q.S., Johnston, A.W.B., Hombrecher, G. and Downie, J.A. (1982) Molecular genetics of mutants of *Rhizobium leguminosarum* which fail to *fix* nitrogen. *Mol. Gen. Genet.* **187**, 166–71.

Martinez, E., Pardo, M.A., Palacios, R. and Cevallos, M.A. (1985) Reiteration of nitrogen fixation gene sequences and specificity of *Rhizobium* in nodulation and nitrogen fixation in *Phaseolus vulgaris*. *J. Gen. Microbiol.* **131**, 1779–86.

Martinez, E., Palacios, R. and Sánchez, F. (1987) Nitrogen-fixing nodules induced by *Agrobacterium tumefaciens* harboring *Rhizobium phaseoli* plasmids. *J. Bacteriol.* **169**, 2828–34.

Moffett, M.L. and Colwell, R.R. (1968) Adansonian analysis of the Rhizobiaceae. *J. Gen. Microbiol.* **51**, 245–66.

Morrison, N.A., Hau, C.Y., Trinick, J.M., Shine, J. and Rolfe, B.G. (1983) Heat curing of a sym plasmid in a fast-growing *Rhizobium* sp. that is able to nodulate legumes and the non-legume *Parasponia* sp. *J. Bacteriol.* **153**, 527–31.

Norel, F., Desnoues, N. and Elmerich, C. (1985) Characterization of DNA sequences homologous to *Klebsiella pneumoniae nif* H, *D, K* and *E* in the tropical *Rhizobium ORS571*. *Mol. Gen. Genet.* **199**, 352–6.

Piñero, D., Martinez, E. and Selander, R.K. (1988) Genetic diversity and relationships among isolates of *Rhizobium leguminosarum* biovar *phaseoli*. *Appl. Environ. Microbiol.*, **54**, 2825–32.

Prakash, R.K. and Atherly, A.G. (1984) Reiteration of genes involved in symbiotic nitrogen fixation by fast-growing *Rhizobium japonicum*. *J. Bacteriol.* **157**, 785–7.

Prakash, R.K., Schilperoort, R.A. and Nuti, M.P. (1981) Large plasmids of fast-growing Rhizobia:homology studies and location of structural nitrogen fixation (*nif*) genes. *J. Bacteriol.* **145**, 1129–36.

Quinto, C., de la Vega, H., Flores, M., Fernández, L., Ballado, T., Soberón-Chávez, G. and Palacios, R. (1982) Reiteration of nitrogen fixation gene sequences in *Rhizobium phaseoli*. *Nature* (*Lond.*) **299**, 724–6.

Quinto, C., de la Vega, H., Flores, M., Leemans, J., Cevallos, M.A., Pardo, M.A., Azpiroz, R., Girard, M.L., Calva, E. and Palacios, R. (1985) Nitrogenase reductase: a functional multigene family in *Rhizobium phaseoli*. *Proc. Natl Acad. Sci. USA*, **82**, 1170–4.

Ramakrishnan, N., Prakash, R.K., Shantaram, S., Duteau, N.M. and Atherly, A.G. (1986) Molecular cloning and expression of *Rhizobium fredii* USDA 193 nodulation genes: extension of host range for nodulation. *J. Bacteriol.* **168**, 1087–95.

Reanney, D.C. (1978) Coupled evolution: adaptive interactions among the genomes of plasmids, viruses and cells. *Int. Rev. Cytol. Suppl.* **8**, 1–68.

Richaume, A., Scott Angle, J. and Sadowsky, M.J. (1989) Influence of soil variables on *in situ* plasmid transfer from *Escherichia coli* to *Rhizobium fredii*. *Appl. Environ. Microbiol.* **55**, 1730–4.

Roberts, G., Leps, W.T., Silver, L.E. and Brill, W.J. (1980) Use of two dimensional polyacrylamide gel electrophoresis to identify and classify *Rhizobium* strains. *Appl. Environ. Microbiol.* **39**, 414–22.

Rosenberg, C., Boistard, P., Denarie, J. and Casse-Delbart, F. (1981) Genes controlling early and late functions in symbiosis are located on a megaplasmid in *Rhizobium meliloti*. *Mol. Gen. Genet.* **184**, 326–33.

Rostas, K., Kondorosi, E., Horvath, B., Simoncsits, A. and Kondorosi, A. (1986) Conservation of extended promoter regions of nodulation genes in *Rhizobium*. *Proc. Natl Acad. Sci. USA* **83**, 1757–61.

Ruvkun, G.B. and Ausubel, F.M. (1980) Interspecies homology of nitrogenase genes. *Proc. Natl Acad. Sci. USA* **77**, 191–5.

Ruvkun, G.B., Long, S.R., Meade, H.M., van den Bos, R.C. and Ausubel, F.M. (1982) ISRm1: a *Rhizobium meliloti* insertion sequence that transposes preferentially into nitrogen fixation genes. *J. Mol. Appl. Genet.* **1**, 405–18.

Sapienza, C. and Doolittle, W.F. (1982) Unusual physical organization of the *Halobacterium* genome. *Nature* (*Lond.*) **295**, 384–9.

Sapienza, C., Rose, M.R. and Doolittle, W.F. (1982) High-frequency genomic rearrangements involving archaebacterial repeat sequence elements. *Nature* (*Lond.*) **299**, 289–92.

Schofield, P.R., Djordjevic, M.A., Rolfe, B.G., Shine, J. and Watson, J.M. (1983) Molecular linkage map of nitrogenase and nodulation genes in *Rhizobium trifolii*. *Mol. Gen. Genet.* **192**, 459–65.

Schofield, P.R., Gibson, A.H., Dudman, W.F. and Watson, J.M. (1987) Evidence for genetic exchange and recombination of *Rhizobium* symbiotic plasmids in a soil population. *Appl. Environ. Microbiol.* **53**, 2942–7.

Scott, K.F., Rolfe, B.G. and Shine, J. (1983) Biological nitrogen fixation: primary structure of the *Rhizobium trifolii* iron protein gene. *DNA* **2**, 149–55.

Selander, R.K. (1985) Protein polymorphism and the genetic structure of natural populations of bacteria. In: Ohta, T. and Aoki, K. (Eds.) *Population Genetics and Molecular Evolution*. Japanese Science Society Press, Tokio/Springer-Verlag, Berlin, pp. 85–106.

Soberón-Chávez, G., Nájera, R., Olivera, H. and Segovia, L. (1986) Genetic rearrangements of a *Rhizobium phaseoli* symbiotic plasmid. *J. Bacteriol.* **167**, 487–91.

Torok, I. and Kondorosi, A. (1981) Nucleotide sequence of the *R. meliloti* nitrogenase reductase (*nif*H) gene. *Nucl. Acids Res.* **9**, 5711–23.

van Brussel. A.A.N., Tak, T., Wetselaar, A., Pees, E. and Wijffelman, C.A. (1982)

Small leguminosae as test plants for nodulation of *Rhizobium leguminosarum* and other rhizobia and *Agrobacteria* harboring a *leguminosarum sym* plasmid. *Plant Sci. Lett.* **27**, 317–25.

Wang, C.L., Beringer, J.E. and Hirsch, P.R. (1986) Host plant effect of hybrids of *Rhizobium leguminosarum* biovars *viceae* and *trifolii*. *J. Gen. Microbiol.* **13**, 2063–70.

Wilson, J.K. (1944) Over five hundred reasons for abandoning the cross-inoculation groups of legumes. *Soil Sci.* **58**, 61–9.

Young, J.P.W. and Wexler, M. (1988) Sym plasmid and chromosomal genotypes are correlated in field populations of *Rhizobium leguminosarum*. *J. Gen. Microbiol.* **134**, 2731–9.

Zurkowski, W. (1982) Molecular mechanism for loss of nodulation properties of *Rhizobium trifolii*. *J. Bacteriol.* **150**, 999–1007.

9 Engineering Microbes for Function and Safety in the Environment

DARYL F. DWYER* and KENNETH N. TIMMIS*
Department of Medical Biochemistry, University of Geneva, Switzerland

* *See pp. x–xi for present address.*

9.1 INTRODUCTION

In recent years, a number of classes of genetically engineered microorganisms (GEMs) have been developed for biotechnological processes that require introduction of the modified organisms into the environment in substantial numbers. This differs from earlier and the majority of current applications of GEMs, which involve GEM containment and ultimate destruction, and has elicited concern about potential risks for the environment. The majority of GEMs do not constitute particularly novel genotypes, nor are they intended for introduction into ecological situations that are unusual for the parental organism. Such GEMs do not therefore differ significantly from traditionally selected microbes with modified properties (e.g. vaccine strains of polio virus and *Salmonella typhi* Ty21a; nitrogen-fixing, plant root nodulated bacteria such as *Rhizobium*) that have been introduced in substantial quantities into the environment for extended periods of time without any untoward effects being detected. Nevertheless, the large number of different GEMs currently under development for introduction into the environment, a few of which may constitute novel genotypes, render dependence upon generalizations unjustified, particularly in view of the limited information available on the fate and ecological consequences of such organisms once introduced into target ecosystems.

There is therefore a need, on the one hand, to avoid irrational concerns that ignore the significant body of case history which indicates a lack of adverse effects on the environment by the vast majority of introduced modified microorganisms and, on the other, to augment substantially our base of relevant data so that we can assess more accurately the likely fate of a wider range of modified microbes. There is also a need to develop

Introduction of Genetically Modified Organisms into the Environment. Edited by Harold A. Mooney and Giorgio Bernardi

strategies to build into particularly novel GEMs, whose fate may be less easily predicted, or into GEMs perceived to represent some degree of risk, containment features that preclude the development of adverse consequences following introduction.

The features of a GEM that are crucial for its effective and safe functioning in the environment include:

1. Its ability to survive and multiply in the ecosystem into which it is introduced
2. The stability of its new genetic material and the potential of this material to transfer horizontally to indigenous organisms
3. The ability of the GEM to carry out effectively the function for which it was designed under the conditions prevailing in the ecosystem
4. The effects, if any, of the GEM on the structure and function of the ecosystem into which it is introduced

9.2 SURVIVAL AND PROPAGATION OF A GEM IN ITS TARGET ECOSYSTEM

9.2.1 GENERALITIES

The most crucial property of a GEM which governs both its efficacy of function and its risk potential is its ability to survive and multiply in the ecosystem into which it is introduced. As a general rule, microbes well adapted to cultivation in the laboratory, through repeated passage in laboratory media, become less robust and persist less well upon return to the natural environment. Similarly, organisms originating from an ecosystem (e.g. topsoil) different from the target ecosystem (e.g. groundwater) are less likely to persist in comparison to those originally derived from the target ecosystem. The behaviour of a GEM constructed by modification of an organism derived from the target ecosystem will be more predictable and will most likely represent a lower risk than that of a GEM constructed from an organism isolated from a different ecosystem. These considerations lead to the conclusion that, other things being equal, selection of an organism from the ecosystem which is to be the target of the modified derivative of the organism, with as little propagation in the laboratory as possible before, during and after its modification, will provide a GEM of greatest efficacy and predictability and of lowest risk.

9.2.2 TECHNIQUES FOR TRACKING GEMs

In order to follow the fate of a GEM in the environment, it is necessary to detect and quantify it in time and space. Both the organism and the genetic information that constitutes the modification must be tracked simultaneously and independently so that both loss of the new information from the GEM and its possible lateral transfer to indigenous microorganisms can be assessed. The principal detection and quantification procedures for monitoring the survival and propagation of a GEM following its introduction into a particular ecosystem are listed in Table 9.1. These differ substantially from one another in specificity and sensitivity. A method exhibiting a high degree of specificity is clearly required for detecting a GEM in an ecosystem containing closely related organisms. On the other hand, a highly sensitive procedure is needed in those cases where persistence of a GEM at very low numbers is important (e.g. in the case of a live vaccine strain whose principal aspect of risk is the possibility of it acquiring greater pathogenic potential). The method(s) used to track a GEM must therefore be selected on the basis of the quality of the information needed. Note that, in general, two procedures must be used in order to track both the organism itself and the newly acquired genetic information (Table 9.1).

Traditional methods for the detection and enumeration of specific microbes generally involve sample dilution and plating for single colonies on solidified selective media. The medium may be selective for a natural property of the

Table 9.1 Methods for tracking GEMs

Component	Phenotype tracking	Genotype tracking
The organism	1. Selective medium for preexisting or newly incorporated metabolic determinant (e.g. lac,[a] nal[b])	1. Nucleic acid probe for a specific natural sequence (e.g. 23S RNA)
	2. Specific antibody to characteristic antigen	2. Nucleic acid probe for a fingerprint sequence[c]
The genetic modification	1. Medium selective or indicative of the modification	1. Nucleic acid probe for introduced DNA sequence
	2. Medium indicative of a closely linked marker (e.g. lac, C230[d])	2. Nucleic acid probe for closely linked fingerprint sequence[c]
	3. Antibody to product of new gene	

[a]Lactose utilization.
[b]Resistance to nalidixic acid.
[c]Rare oligonucleotide sequence introduced into the GEM for the purpose of being able to identify it or the new information incorporated during its construction.
[d]Catechol 2,3-dioxygenase, which causes colonies to turn yellow when sprayed with catechol as a result of its transforming the latter to 2-hydroxymuconic semialdehyde.

organism or for a newly acquired property (e.g. lactose utilization, resistance to nalidixic acid) that has been introduced specifically for the purpose of tracking the GEM (Table 9.1) and which differs from the introduced property that constitutes the crucial new functional aspect of the GEM. Exceptionally, this latter property may also serve as a basis for specific selection or detection of the GEM on solid media (Rojo *et al.*, 1987), thereby enabling both the organism and the newly acquired genetic information to be independently tracked by plating techniques (D.F. Dwyer, S.W. Hooper, F. Rojo, and K.N. Timmis, submitted for publication). Where the newly acquired information does not itself confer a phenotypic property detectable by plating procedures, it can be directly linked to a marker that does (e.g. genes encoding lactose utilization (Barry, 1986; Drahos *et al.*, 1986), catechol 2,3-dioxygenase, etc. (Mermod *et al.*, 1986a)).

Despite the simplicity of plating procedures, they do have a number of limitations in that they not only require the organism to be efficiently recoverable from environmental samples but also to be able to grow and develop colonies on the selective medium. Many organisms present in natural environments do not form colonies when directly plated on standard laboratory media, let alone strongly selective media, and require special resuscitation media and growth conditions in order to propagate themselves in the laboratory (Palmer *et al.*, 1984; Lechevallier *et al.*, 1987). It is also possible that a GEM may not express the genes whose functions are necessary for survival on the selective medium employed or that these genes may be lost while the GEM is in the environment.

The problems of recoverability of organisms from natural environments and their growth on laboratory media can be avoided by direct examination of samples or extracts of samples using an antibody specific for an antigen produced by the GEM. Direct enumeration of microbes may be effected by microscopic examination using fluorescent antibodies directed against a surface antigen (Bohlool and Schmidt, 1980) or indirect quantitation may be carried out on extracts of surface or non-surface antigens using standard immunological procedures. If monoclonal antibodies are employed, this method can have both high sensitivity and high specificity. Since GEMs constructed for introduction into the environment will generally express the newly acquired genetic information, immunological methods can be used to track the organism and the new information simultaneously. Note, however, that precise and reproducible results require a uniform level of synthesis of the marker antigens (and exposure on the cell surface in the case of surface antigens) under all environmental conditions encountered, and this may not always be the case.

The problem of variable expression of marker antigens is obviated by the use of nucleic acid hybridization procedures in conjunction with oligonucleotide probes for homologous sequences in the GEM. Oligonucleotide probes can have homology to a specific DNA sequence of the organism (e.g. to a specific sequence of the 23S RNA gene) or to the added genetic information. Alternatively, it can have homology to a short, rare oligonucleotide (fingerprint) introduced into the chromosome or into (or closely linked to) the newly acquired information, specifically for the purpose of tracking. Nucleic acid probes can be used in various ways. Colony hybridization methods (Grunstein and Hogness, 1975) involve the growth of bacteria isolated from environmental samples on laboratory media and the scoring of these colonies with the probe. This procedure has been shown to be fairly sensitive for the detection of specific bacteria added to environmental samples (Sayler *et al.*, 1985) and for detecting the presence of specific genes in bacterial communities (Barkay *et al.*, 1985). However, it requires the successful isolation of bacteria and their growth in the laboratory. Alternatively, probes can be used to analyse bacterial DNA directly isolated from environmental samples (Kuritza and Salyers, 1985). This procedure has no requirement for the prior laboratory growth of bacteria carrying the gene of interest and allows for quantification of particular genes. In this way, the size of the bacterial population containing the gene can be estimated. When used together the two methods can give a good estimate of the size of the population harbouring the homologous genes.

9.3 STABILITY OF ACQUIRED HETEROLOGOUS DNA AND ITS TRANSMISSION TO INDIGENOUS ORGANISMS

9.3.1 INSTABILITY OF STRUCTURE AND INHERITANCE OF CLONED DNA

Many hybrid plasmids tend not to be inherited stably in populations of actively dividing bacteria in the absence of continuous selection conditions that kill plasmid-free bacteria. There are several reasons for this (Table 9.2), including disequilibrations in cellular metabolism and physiology or the perturbation of membrane structure and function (in the case of cloned

Table 9.2 Some causes of instability of DNA sequences cloned in plasmid vectors

1. Determinants on insert that adversely affect vector maintenance (e.g. strong promoters that cause excessive transcription into vector replication determinants)
2. Determinants on insert that adversely affect normal cell functioning (e.g. that encode products toxic for host cell)
3. Presence on insert of recombinationally aggressive sequences (e.g. insertion element) or of sequences already present in the vector (that create duplications).

genes of membrane proteins), resulting from the synthesis of a heterologous protein or proteins (or their products), or the higher-than-normal rate of synthesis of a heterologous or homologous protein (Mermod *et al.*, 1986a). The perturbation in optimal cellular functioning caused by a hybrid plasmid constitutes a selection pressure for elimination of the cloned gene. This may occur by insertional inactivation of the gene, deletion of the gene through recombination between repeated sequences bracketing the cloned sequence (or part of it) or through the activity of deletogenic sequences in the neighbourhood of the cloned gene, or by loss of the hybrid plasmid. Which of these processes is dominant in any situation depends upon the nature of sequences in the hybrid plasmid, the properties of the host strain, and particularly the inherent instability of the hybrid plasmid in growing cells. The molecular basis of instability of an acquired property can often be determined precisely by structural analysis of the cloned genes in individual clones or in mixed populations through the Southern hybridization procedure (Southern, 1975; Meinkoth and Wahl, 1984), using the cloned gene as a probe.

9.3.2 STRATEGIES TO INCREASE STABILITY: INSERTION INTO THE CHROMOSOME

The main causes of instability of a hybrid plasmid, namely the high-level expression of cloned genes and inherent instability of many current cloning vectors, may be addressed in several ways. Whereas high-level expression may result from the existence of an unnatural constitutive or partially constitutive promoter upstream of and oriented towards the cloned gene, it often reflects a gene dosage effect of DNA sequences cloned into high copy number vectors, such as pBR322 and pUC derivatives. The cloning of genes together with their native promoters and natural regulatory determinants into low copy number vectors such as pRK290 (Ditta *et al.*, 1980) or pLG339 (Stoker *et al.*, 1982) generally produces hybrid plasmids with improved stability. The best stability is, however, obtained in most cases through eliminating altogether the gene dosage phenomenon, i.e. through reducing the copy number of the gene to one per chromosome, the natural level for many genes, by inserting the gene into the chromosome itself (Rojo *et al.*, 1987). This generally has three important positive consequences: (1) reduction in expression of the cloned gene to its normal level; (2) insertion of the gene into an essential replicon that cannot by definition be lost from a viable cell; and (3) reduction in the tendency of the gene to transfer laterally to other organisms (see below).

There are basically two strategies for inserting cloned genes into host chromosomes. The most versatile strategy is to insert the genes into a transposon present on a suitable delivery system and then to cause transposition of the hybrid transposon into the chromosome of the selected host strain (Simon *et al.*, 1983; Tait *et al.*, 1983). Typically, the transposon into which the gene of interest is cloned is located on a plasmid that can be propagated stably in one host (e.g. *Escherichia coli*) but not in target bacteria (e.g. *Pseudomonas putida*) but that can conjugally transfer at high frequency to the latter (Mermod *et al.*, 1986a). The hybrid plasmid is then transferred by conjugation to the target organism, and transconjugants expressing a marker of the transposon are selected. Since the plasmid itself cannot survive in the target organism, the only means by which the transposon marker can persist is through transposition of the hybrid transposon from the donor plasmid to a stable replicon (e.g. the chromosome) of the recipient. Although some derivatives obtained in this way will contain the hybrid transposon in an important gene, and will thus be compromized in their metabolism and growth, others will not and will be healthy hybrids. The versatility of this approach is that a single transposon can be used for many different target bacteria. However, much development of transposon cloning vectors remains to be done.

The other strategy is to clone the gene into a previously cloned segment of the chromosome of the host to be used and then to affect targeted insertion of the gene by recombination between the regions of homology on the chromosome and the introduced hybrid plasmid (Harayama and Don, 1985). Since the site for insertion on the chromosome is selected for an absence of deleterious effects of the manipulations, all targeted insertions obtained should be healthy. However, with this strategy a new cloning vector must be constructed essentially each time a different host is used.

9.3.3 HORIZONTAL TRANSFER OF CLONED DNA

A critical aspect of a GEM is whether it horizontally transfers its new information to indigenous microorganisms and, if so, to which members and at what frequency. Non-GEMs transfer DNA in soil culture (Graham and Istock, 1978; Weinberg and Stotzky, 1972) and thus the potential for transfer by GEMs in nature exists. It should be emphasized that horizontal transfer of such information is not necessarily undesirable because it may lead to the evolution of organisms that carry out the desired activity better than the original GEM. However, since horizontal transfer in the environment cannot be controlled and its consequences may not be predictable *a priori*, effort should be made to reduce it to a minimum until sufficient data are obtained to permit valid predictions.

A number of current cloning vectors, particularly broad host range vectors that are used extensively for genetic engineering of soil bacteria, are readily transmissible when present in a cell concommitantly carrying a conjugative plasmid such as F, ColI, RP1/RP4/RK2, etc., because they retain the mobilization functions (*mob* and *oriT*) of the original replicon from which they were constructed (Bagdasarian *et al.*, 1979; Bagdasarian and Timmis,

1982). In the case of the pKT series of vectors, transfer-deficient (Mob$^-$) derivatives are available that, in combination with *Pseudomonas putida* host strain KT2440, are certified by the Office of Recombinant DNA Activities, National Institutes of Health, USA, as the cloning vectors of an HV1 (host-vector 1) system (Bagdasarian *et al.*, 1981). However, even with these and with Mob$^-$ HV1 vectors of *E. coli* such as pBR322, low-frequency conjugal transfer can be demonstrated under laboratory conditions (Guyer, 1978). There is, however, only limited information about the ability of either hybrid plasmids engineered for a particular function in the environment or naturally occuring plasmids to transfer to indigenous organisms in their target environment or in model ecosystems that closely mirror target environments (Gowland and Slater, 1984; Gealt *et al.*, 1985; Mancini *et al.*, 1987).

Horizontal transfer can also occur by transduction, following infection of the donor by a virus, or by transformation, following death and lysis of the donor organisms. Again there is little information available on frequencies of transfer of specifically engineered (i.e. non-model) plasmids or natural plasmids by these processes in natural environments or model ecosystems (Saye *et al.*, 1987). It seems, however, likely that transfer frequencies under such conditions will reflect both the copy number of the new gene and the ability of it either to become a relatively small replicon capable of independent propagation (i.e. a plasmid) or to become integrated into an existing replicon (e.g. the chromosome) in the new host. It is therefore to be expected that, all things being equal, a gene on a high copy number plasmid will be transferred at a higher frequency than one on a low copy number plasmid, and that a gene integrated in the chromosome will be transmitted at the lowest frequency, both because it is present as a single copy and because it constitutes only a tiny part of a very large replicon that in all probability will be extremely inefficient in transferring to a new organism and integrating into its chromosome.

9.3.4 STRATEGIES TO DIMINISH HORIZONTAL TRANSFER OF DNA FROM GEMS

Although the institution of additional barriers to horizontal transfer has not been extensively explored, it is quite conceivable that this would be readily achievable (Table 9.3). In the case of the conjugal transfer of bacterial DNA, although it is known that most of the functions needed for conjugation and mobilization are plasmid-encoded, host mutants have been described that are blocked in conjugal transfer (Willetts and Wilkins, 1984). The introduction of such defects into GEMs carrying conjugation- and mobilization-deficient vectors, and the analysis of transfer frequencies from such bacteria with combined transfer defects, would seem well worth exploring. Moreover, it is not beyond the realms of current capabilities to take a GEM system exhibiting low transfer rates (by any means) and to engineer it in

Table 9.3 Horizontal transfer of cloned genes and possible means of prevention

Transfer mechanism	Requirements		Experimentally reduced by
	In donor	In recipient	
Mobilization by conjugative plasmid	1. Presence of a conjugative plasmid 2. Origin of transfer 3. Mobilization functions 4. Host functions	Recipient functions	1. Inactivation of mobilization functions 2. Inactivation of host functions (donor) 3. Linkage of a 'kill' gene to cloned gene
Transduction	Transducing phage	Sensitivity to transducing phage	Linkage of a 'kill' gene to cloned gene
Trans-formation	None	Ability to take up exogenous DNA	1. Production of active nucleases by donor 2. Linkage of a 'kill' gene to cloned gene

such a way that any recipient of the modified DNA segment would rapidly die. There are a number of genes (e.g. genes of restriction endonucleases, colicins, etc.) that, when expressed in the absence of another gene (e.g. genes of modification enzymes, colicin immunity proteins, etc.), cause killing of the cell (e.g. see Kuhn *et al.*, 1986). If the 'kill' gene is inserted close to the DNA fragment in question and the 'protection' gene inserted at a site far from this fragment or, preferably, into another replicon in the cell, then the GEM itself would be viable whereas a recipient of the 'kill' gene would be killed because co-acquisition of the 'protection' gene in a low transfer system would be exceptionally low. The potential utility of a variation on this theme, involving the plasmid Rl *hok* gene, was recently demonstrated (Molin *et al.*, 1987). Note that engineering a GEM to produce an active and stable nuclease may greatly diminish the transforming ability of its DNA released after cell death (e.g. see Timmis and Winkler, 1973).

9.3.5 ASSESSMENT OF RATES OF HORIZONTAL TRANSFER AND DESTINY OF TRANSFERRED DNA

The assessment of horizontal transfer of cloned genes can be accomplished with a variety of approaches. The *potential* for transfer under different conditions can be studied conveniently in the laboratory. Conjugal transfer of the new information of the GEM to selected recipients using filter matings and strongly selective conditions for the isolation of transconjugants enables detection of low transfer frequencies under optimal conditions. Similarly, the influence on low transfer frequencies of a third partner carrying a promiscuous conjugative plasmid can be examined using triparental matings. Transfer frequencies between pairs and triples of bacteria on more natural surfaces (sand, clay, etc.) can be analysed following adsorption to or biofilm formation on such surfaces by the partners. The potential influence on transfer rates of nutrients, temperature, pH, ionic strength, the presence of noxious compounds, etc., can all be assessed in such types of experiments.

The isolation of phages able to grow on the donor strain enables assessment of the potential for transductional transfer of the new information to selected recipients. The isolation of total DNA and its application to selected recipients present in different media and metabolic states provides information on the potential for the transfer of new information by transformation.

Actual transfer rates in natural environments are less readily assessed because conditions are not easily controlled and because all non-GEM bacteria present in a target ecosystem, most of which cannot be individually selected, are potential recipients. Simultaneous addition of the GEM and a target recipient to an ecosystem, with assessment only of target recipients having acquired the new information, is a means of obtaining precise data, but this information may have limited significance.

Model ecosystems offer a useful intermediate stage between a natural ecosystem and laboratory experiments, in that conditions can be more carefully controlled and the system can be fully contained, thus avoiding the possibility of introducing GEMs into the environment before they or their potential effects are sufficiently characterized (Sagik *et al.*, 1981).

Procedures available to detect and measure gene transfer range from the *direct selection* of transconjugants using selection markers of the recipients and of the transferred DNA to *indirect detection* of transconjugants using nucleic acid and/or antibody probes and/or plating techniques to separately detect the donor and the new genetic information, and thereby to identify non-donor organisms carrying the new information. Whereas the direct selection of transconjugants detects low transfer frequencies, indirect detection systems are only able to reveal high transfer frequencies because of the presence of donor bacteria. Since donor bacteria need not be present in experiments designed to evaluate transduction (application of phages that were propagated on the donor) and transformation (application of donor DNA), detection systems here can be more sensitive because of the lower background. However, the absence of donor bacteria is unnatural and may negatively influence transduction and transformation frequencies, particularly if proximity of donor and recipient (as is presumably the case with bacteria absorbed to soil particles or growing in biofilms) is important for such mechanisms of DNA transfer.

9.4 GEM FUNCTIONING

In order to be useful, a GEM must survive for a minimal period of time and express its new genetic information (and most probably a number of other non-modified key functions) for at least this period of time and at some minimal level, in the target environment in order to successfully accomplish the job for which it was designed. In this regard, it is encouraging that aromatic-degrading bacteria which have either been isolated from activated sewage sludge (Focht and Shelton, 1987) or which have modified genotypes obtained by conjugative plasmid transfer (Jain *et al.*, 1987) both survive and function as aromatic degraders upon introduction into soil and groundwater microcosms.

Clearly there are a large number of variables that will influence these properties of the GEM and that must be evaluated. The most obvious of these is the ability of the GEM to survive in its target environment, and parameters important for this have been alluded to above. Similarly, the stability of the modified DNA in the GEM is of crucial importance and some strategies to optimize stability, including retention of native gene regulation systems, have been considered (Table 9.4).

For many applications, there will be no specific environmental signal that can be used to switch on expression of cloned genes when the engineered function is needed and to switch them off when not needed. Here, the cloned genes will need to be expressed constitutively and this may represent an energetic (and perhaps a physiological) burden on the GEM that will reduce its survival or the survival of the modified genetic information (see above). In other applications, however, it will be possible and preferable to retain native regulatory circuits that do respond to appropriate environmental

Table 9.4 Engineering GEMs for function

1. Stability of introduced genetic information
   (a) Incorporate genes into the chromosome
   (b) Postsegregational killing of plasmid-free cells[a]

2. Control of expression
   (a) Specific regulators
   (b) Back-up regulation or low-level constitutive expression for conditions of low inducer concentration

3. Accessory functions
   (a) Colonization mechanisms
   (b) Chemotactic functions
   (c) Improved NADH generation
   (d) Predator protection mechanisms

[a]Only effective if the GEM multiples well in the environment.

signals: e.g. the expression of adhesion antigens and other virulence-associated factors at 37°C (body temperature; e.g. see Timmis and Manning, 1986) by live vaccines and the expression of catabolic genes in response to the presence of specific pollutants in microbes designed to degrade such pollutants (Ramos *et al.*, 1987). In such cases, the lack of or low-level expression of the cloned genes when their products are not required will diminish their energetic and physiological burden on the GEM and will have less effect on its survival.

The use of microbes to degrade chemicals toxic at the p.p.b. level poses a particular problem because organic chemicals typically induce synthesis of enzymes for their catabolism when present only in p.p.m. concentrations (Mermod *et al.*, 1986b). Thus, catabolic genes whose expression is controlled by native regulatory circuits will stop being expressed and the desired activity of the GEM will cease to function when the level of the pollutant drops to a level that still constitutes a significant environmental hazard. Maintenance of the synthesis of catabolic enzymes at p.p.b. levels of the pollutant can be accomplished through having constitutive synthesis of the catabolic enzymes, but since catabolic pathways typically involve many enzymes, this would constitute a considerable drain upon the energy resources of the cell. It is therefore preferable to maintain the native regulatory circuits but to engineer into the system a back-up expression circuit that would switch on low-level synthesis of the catabolic enzymes when the cell finds itself in substrate-poor environments, i.e. in response to a starvation or stress signal (Table 9.4). Further means of promoting elimination of pollutants present at low levels include the incorporation or engineering of high-affinity transport proteins for certain compounds and the manipulation of chemotactic responses to accomplish directional movement of bacteria towards substrates adsorbed to particulate matter.

GEM functioning (Table 9.4) might also be improved by the incorporation of properties that permit specific adhesion to surfaces upon which the organism would function optimally or that provide protection from certain predators active in the target environment. Where the function of the GEM involves metabolism of a particular substrate that causes imbalances in parameters like NADH production, engineering of these secondary parameters to compensate the imbalances may be useful if feasible (Ingram *et al.*, 1987).

9.5 EFFECT OF THE GEM ON STRUCTURE AND FUNCTION OF TARGET ECOSYSTEM

At present we have rather limited information on microbial ecology and about microbial ecosystem structure and function. In some, particularly anaerobic ones, a significant proportion of the bacteria present have neither been satisfactorily classified nor named, nor are they readily isolated by standard procedures. It is therefore rather difficult at present to examine effects on the structure and function of some ecosystems provoked by introduction of a GEM. However, the techniques necessary for studying microbial ecosystems are now available and we can anticipate great progress in this area over the next few years. Moreover, a bank of data on the interactions of microbes with their environment and with each other is growing (Liang *et al.*, 1982; Stotzky and Krasovsky, 1981; Stotzky and Babich, 1984) and may soon enable concepts and mathematical models of microbial interactions to be formulated. Until this time, studies on the influence of introduced GEMs on ecosystem structure and function will rely primarily on measurement of transfer of modified DNA from the GEM to indigenous organisms, using procedures outlined above, indirect assessment of population complexity, and analysis of general aspects of community functioning.

9.5.1 METHODS TO FOLLOW THE FATE OF INDIVIDUAL MICROORGANISMS IN AN ECOSYSTEM

In simple communities consisting of only a few distinct species, all of which can be isolated and propagated in pure culture, it will be relatively straightforward to generate specific antibody or nucleic acid probes that will permit each member to be quantified in time and space. Even simpler perhaps will be the use of pulsed field electrophoresis (Schwartz and Cantor, 1984) and contour clamped homogeneous field electrophoresis (Chu *et al.*, 1986) to analyse total DNA extracted from samples and cut with an infrequent cutting restriction endonuclease. A knowledge of the fragmentation pattern of individual members may permit quantification of all members from the patterns obtained with total DNA extracts. This method should even be applicable to larger communities if combined with nucleic acid hybridization using a mixture of probes, each one specific for one of the members. Another simple analytical procedure that may be useful is gas chromatography (GC) of fatty acids (Smith *et al.*, 1986). Different species contain different fatty acids that appear as a distinct GC profile. Provided that the community is not complex, GC analysis of total fatty acids will be a rapid means of quantifying all members. Another approach is to identify certain bacteria by determining the sequences of unique segments of 5S rRNA (MacDonell and Colwell, 1984).

In more complex communities, higher resolution is required and the preparation of individual reagents specific to each member may be impractical. In such cases, which are likely to be the majority, indirect assessment of community complexity and the individual tracking of uncharacterized organisms by means of probes may be necessary. One promising approach is the raising of polyclonal antibodies against cell surface antigens of entire microbial populations, and their use to analyse by Western blotting (Towbin *et al.*, 1979) total proteins extracted from community samples and fractionated on two-dimensional polyacrylamide gels (O'Farrell, 1975; Neidhardt *et al.*, 1983). Subsequent probing of such blots with antibodies to one or several specific antigens would provide additional information about subpopulations in the community.

9.5.2 ASSESSMENT OF ECOSYSTEM FUNCTIONING

Although risk assessment related to environmental introduction of modified microorganisms has understandably until now focused largely upon risks for the health of humans, animals, and crop plants, it must increasingly take into account risks for microbial ecosystem structure and function (Marx, 1987). A GEM that alters the structure and/or function of an ecosystem into which it is introduced might do so either as a result of its activities unrelated to the genetic modification, an activity related to the genetic modification, or behaviour resulting from interplay of the two.

To predict the impact of a GEM upon a given ecosystem requires knowledge about the reciprocal effects of the GEM and its environment and of the physical and biological parameters that control certain activities (Pritchard *et al.*, 1987). Ecological factors such as temperature, soil type, pH, nutrient and moisture availability, competition, and predation by indigenous microorganisms, etc. (Stotzky and Babich, 1984; Stotzky and Krasovsky, 1981), which affect survival and activity as well as the stability and expression of introduced genetic information of GEMs and indigenous microorganisms, need to be delineated. In certain cases it will be possible to assess direct competitive effects, e.g. the exclusion of epiphytic bacteria by introduced Ice$^-$ mutants of *Pseudomonas syringae* (Lindow, 1987). In others, however, possible effects may only be detected indirectly by measurement of alterations in global ecosystem processes such as energy flow and nutrient cycling (Levin and Harwell, 1986).

Determination of a GEM's possible effect on an ecosystem, as well as its ability to function effectively therein, may be conveniently ascertained using model systems. A data base covering the interplay of model GEMs and various types of model ecosystems would provide: (1) a predictive means for evaluating whether introduction of a given GEM into the environment would be accompanied by significant risk, (2) guidelines for constructing GEMs to meet certain environmental standards for activity and survival, and (3) methods for determining the fate of a GEM and of its modified genetic component(s) following environmental introduction (Levin *et al.*, 1987).

9.6 ENGINEERING GEMs FOR ENVIRONMENTAL SAFETY

Two aspects relating to the safety of GEMs to be introduced into the environment are the risk that exists because sufficient information on the behaviour of the GEM in the target ecosystem is lacking and the more serious risk that is associated with the use of GEMs derived from microbes that exhibit a known hazard, such as live vaccine strains derived from pathogenic microorganisms. In the first instance, prudence indicates that the new genetic information should not be able to transfer laterally at significant frequencies and the GEM should not be able to propagate itself at a rate that allows it to significantly disturb the normal functioning of the ecosystem or to displace any members of the community. In the second, the GEM must additionally be modified so that it is unable to regain its prior hazard potential and so that it has only limited capacity to survive in its target ecosystem.

Currently available information suggests that most GEMs will not be as competitive as indigenous members of an ecosystem, will tend to be maintained at low levels, and thus will not have a significant impact on the ecosystem structure or function. This appears to be the case for *Pseudomonas* sp. B13 derivatives able to degrade chlorinated benzoates and phenols that have been studied in a model activated sludge ecosystem (D.F. Dwyer, S.W. Hooper, F. Rojo, and K.N. Timmis, submitted for publication). When such bacteria are introduced into this ecosystem at a level equal to that of all other bacteria in the ecosystem able to form colonies on laboratory media, their numbers quickly fall to 0.1–0.01% of this level and then stabilize. These numbers only increase significantly if the ecosystem is placed under strong selection pressure for the GEM function and even then the increase is modest and not maintained. The GEM nevertheless functions in the way for which it was designed. Thus, as indicated above, an organism taken from the target environment and exhibiting no pathogenic potential for animals or plants, and modified in the laboratory to perform some specific task, is unlikely to provoke changes in the community structure and function when reintroduced into its environment. This is of course provided that steps are taken to minimize lateral transfer of its modified DNA to indigenous organisms.

If the host strain is judged to be safe but there seems to be a risk associated with the modified or new DNA, then the disadvantages of having the DNA on a plasmid vector, in terms of instability of the GEM, can constitute advantages in terms of safety (Table 9.5). In this case, vector plasmid instability can be increased further by deleting DNA sequences (*par*) necessary for precise partitioning of the vector between daughter cells at cell division (e.g. see Cohen *et al.*, 1986).

If the GEM itself is judged to represent a significant risk there are two

Table 9.5 Engineering for safety/containment

1. Unstable inheritance of introduced genes (unstable plasmid, par^-)
2. Poor host survival (auxotroph, recA, defective in global regulation of metabolism)
3. Autodestruct mechanism (e.g. $restriction^+$, $bacterocin^+$)

general strategies to diminish its risk. One is to reduce its survival capacity such that it can survive for only a limited period of time outside of the laboratory. For example, strains of *Salmonella* currently under development as live vaccines have been attenuated by the introduction of non-reverting mutations that inactivate either the ability to synthesize aromatic amino acids (Levine *et al.*, 1987) or the cAMP–CRP global regulatory circuit of the cell (Curtiss and Kelly, 1987). Other global circuits that might be good targets for reducing the survival of GEMs include the RecA recombination system and the Rel stringent response.

A second approach is to engineer the GEM in such a way that it dies rapidly once its specific function is no longer required. The construction of such an autodestruct GEM is possible where expression of the function is regulated by an environmental signal, such as the presence of a pollutant, a temperature of 37°C, etc. One strategy here would be to take a bicomponent 'kill–protection' system (e.g. restriction–modification; bacteriocin–bacteriocin immunity) and to arrange for the 'kill' gene to be expressed constitutively and the 'protection' gene to be expressed from the same regulated promoter that is responsible for expression of the GEM function and whose activity is modulated by an environmental signal. For example, if the protection gene in a vaccine strain is expressed from a promoter that is active at 37°C but not at temperatures below 30°C, then the GEM will be viable while in the individual being vaccinated but not when it passes out of the body, where it is exposed to lower temperatures and expresses only the kill gene (see also the use of the *hok* system; Molin *et al.*, 1987).

9.7 CONCLUDING REMARKS

The introduction of GEMs into the environment carries with it not only the promise of health, ecological and economic benefits, but also the responsibility to protect the environment from any potential adverse consequences. It is therefore essential to devise strategies and develop methods for obtaining qualitative and quantitative data concerning survival, competitiveness, etc., of GEMs and of their ability to function in the ecosystems into which they are introduced.

In this overview of experimental approaches to construct safe and effective GEMs for environmental applications, we have attempted to point out those criteria that in our view are important for making strategic choices. On the one hand, we have been dogmatic where we feel a particular choice is dictated. On the other, we have in some instances been highly speculative because new approaches and new applications of developing technologies are clearly needed and we do not yet know with any certainty which of these will ultimately turn out to be most useful.

What is apparent is that, although much basic information is urgently needed about microbial ecology and about the possible influence on the structure and function of an ecosystem into which a GEM is introduced, the experimental techniques to obtain such information are now available or can readily be developed. The application of modern molecular biological and immunological techniques, as well as analytical instrumentation for GC, high-pressure liquid chromatography, mass spectrometry, nuclear magnetic resonance, etc., will undoubtedly revolutionize microbial ecology during the next few years and lead to significant advances in our understanding of fundamental processes of ecosystem functioning and of the important attributes desirable in a GEM destined to be introduced into the environment.

REFERENCES

Bagdasarian, M. and Timmis, K.N. (1982) Host:vector systems for gene cloning in *Pseudomonas*. In: Henle, W., Hofschneider, P.H., Koprowski, H., Melchers, F., Rott, R., Schweiger, H.G. and Vogt, P.K. (Eds.) *Current Topics in Microbiology and Immunology*, Vol. 96, Springer-Verlag, Berlin, pp. 47–67.

Bagdasarian, M., Bagdasarian, M.M., Coleman, S. and Timmis, K.N. (1979) New vector plasmids for gene cloning in *Pseudomonas*. In Timmis, K.N. and Pühler, A. (Eds.) *Plasmids of Medical, Environmental and Commercial Importance*, Elsevier/North Holland, Amsterdam, pp. 411–22.

Bagdasarian, M., Lurz, R., Rückert, B., Franklin, F.C.H., Bagdasarian, M.M., Frey, J. and Timmis, K.N. (1981). Specific purpose cloning vectors. II. Broad host range, high copy number, RSF1010-derived vectors, and a host:vector system for gene cloning in *Pseudomonas*. *Gene* **16**, 236–47.

Barkay, T., Fouts, D.L. and Olson, B.H. (1985) Preparation of a DNA gene probe for detection of mercury resistance genes in gram-negative bacterial communities. *Appl. Environ. Microbiol.* **49**, 686–92.

Barry, G.F. (1986). Permanent insertion of foreign genes into the chromosomes of soil bacteria. *Bio/Technology* **4**, 446–9.

Bohlool, B.B. and Schmidt, E.L. (1980) The immunofluorescence approach in microbial ecology. *Adv. Microbial. Ecol.* **4**, 203.

Chu, G., Vollrath, D. and Davis, R.W. (1986) Separation of large DNA molecules by contour-clamped homogeneous electric fields. *Science* **234**, 1582–5.

Cohen, S.N., Biek, D.P. and Miller, C.A. (1986). Chromosomal and extrachromosomal functions that affect plasmid stability in *E. coli*. In:Levy, S.B. and Novick, R.P. (Eds.) *Banbury Report 24: Antibiotic Resistance Genes: Ecology, Transfer, and Expression*, Cold Spring Harbor Laboratory, Cold Spring Harbor, New York, pp. 247–61.

Curtiss, R. III and Kelly, S.M. (1987) *Salmonella typhimurium* deletion mutants lacking adenylate cyclase and cyclic AMP receptor protein are avirulent and immunogenic. *Infect. Immum.* **55**, 3035–43.

Ditta, G., Stanfields, S., Corbin, D. and Helinski, D.R. (1980) Broad host range DNA cloning system for gram-negative bacteria: construction of a gene bank of *Rhizobium meliloti. Proc. Natl Acad. Sci. USA.* **77**, 7347–51.

Drahos, D.J., Hemming, B.C. and McPersons, D. (1986) Tracking recombinant organisms in the environment: B-galactosidase as a selectable non-antibiotic marker for fluorescent *Pseudomonas. Bio/Technology* **4**, 439–44.

Focht, D.D. and Shelton, D. (1987). Growth kinetics of *Pseudomonas alcaligenes* C-O relative to inoculation and 3-chlorobenzoate metabolism in soil. *Appl. Environ. Microbiol.* **53**, 1846–9.

Gealt, M.A., Chai, M.D., Alpert, K.B. and Boyer, J.C. (1985) Transfer of plasmids pBR322 and pBR325 in wastewater from laboratory strains of *Escherichia coli* to bacteria indigenous to the waste disposal system. *Appl. Environ. Microbiol.* **49**, 836–41.

Gowland, P.C. and Slater, J.H. (1984) Transfer and stability of drug resistance plasmids in *Escherichia coli* K12. *Microbiol. Ecol.* **10**, 1–13.

Graham, J.B. and Istock, C.A. (1978) Gene exchange and natural selection cause *Bacillus subtilis* to evolve in soil culture. *Science* **204**, 637–9.

Grunstein, M. and Hogness, D.S. (1975) Colony hybridization: a method for the isolation of cloned DNAs that contain a specific gene. *Proc. Natl Acad. Sci. USA* **72**, 3961–5.

Guyer, M.S. (1978) The γ δ sequence of F is an insertion sequence. *J. Molec. Biol.* **126**, 347–65.

Harayama, S. and Don. R.H. (1985) Catabolic plasmids: their analysis and utilization in the manipulation of bacterial metabolic activities. In Setlow, J.K. and Hollaender, A. (Eds.) *Genetic Engineering*, Vol. 7, Plenum Press, New York, pp. 283–307.

Ingram, L.O., Conway, T., Clark, D.P., Sewell, G.W. and Preston, J.F. (1987) Genetic engineering of ethanol production in *Escherichia coli. Appl. Environ. Microbiol.* **53**, 2420–5.

Jain, R.K., Sayler, G.S., Wilson, J.T., Houston, L. and Pacia, D. (1987) Maintenance and stability of introduced genotypes in groundwater aquifer material. *Appl. Environ. Microbiol.* **53**, 996–1002.

Kuhn I., Stephenson, F.H., Boyer, H.W. and Greene, P.J. (1986) Positive-selection vectors utilizing lethality of the *Eco* R1 endonuclease. *Gene* **44**, 253–63.

Kuritza, A.P. and A.A. Salyers (1985) Use of a species-specific DNA hybridization probe for enumerating *Bacteroides vulgatus* in human feces. *Appl. Environ. Microbiol.* **50**, 958–64.

Lechevallier, M.W., Camper, A.K., Broadaway, S.C., Henson, J.M. and McFeters, G.A. (1987) Sensitivity of genetically engineered organisms to selective media. *Appl. Environ. Microbiol.*, **53**, 606–9.

Levin, S.A. and Harwell, M.A. (1986) Potential ecological consequences of genetically engineered organisms. *Environ. Managem.* **10**, 495–513.

Levin, M.A., Seidler, R., Borquin, A.W., Fowle, J.R. and Barkay, T. (1987) EPA developing methods to assess environmental release. *Bio/Technology* **5**, 38–45.

Levine, M.M., Herrington, D., Morris, J.G., Lasonsky, G., Murphy, J., Tall, B. and Stocker, B. (1987) Safety, inactivity, immunogenicity, and *in vivo* stability of two attenuated auxotrophic mutant strains of *Salmonella typhi* 541Ty and 543Ty, used as oral vaccines in man. *J. Clin. Invest.* **79**, 888–902.

Liang, L.N., Sinclair, J.L., Mallory, L.M. and Alexander, M. (1982) Fate in model ecosystems of microbial species of potential use in genetic engineering. *Appl. Environ. Microbiol.* **44**, 708–14.

Lindow, S.E. (1987) Competitive exclusion of ephiphytic bacteria by Ice-*Pseudomonas syringae* mutants. *Appl. Environ. Microbiol.* **53**, 2520–7.

MacDonell, M.T. and Colwell, R.R. (1984) Identical 5S rRNA nucleotide sequence of *Vibrio cholerae* strains representing temporal, geographical, and ecological diversity. *Appl. Environ. Microbiol.* **48**, 119–21.

Mancini, P., Fertels, S., Nave, D. and Gealt, M.A. (1987) Mobilization of plasmid pHSV106 from *Escherichia coli* HB101 in a laboratory-scale waste treatment facility. *Appl. Environ. Microbiol.* **53**, 665–71.

Marx, J.L. (1987) Assessing the risks of microbial release. *Science* **237**, 1413–17.

Meinkoth, J. and Wahl, G. (1984) Hybridization of nucleic acids immobilized on solid supports. *Analyt. Biochem.* **138**, 267–84.

Mermod, N., Lehrbach, P.R., Don, R.H. and Timmis, K.N. (1986a) Gene cloning and manipulation in *Pseudomonas.* In: Sokatch, J.R. and Ornston, L.N. (Eds.) *The Bacteria*, Academic Press, New York, pp. 325–55.

Mermod, N., Ramos, J.L., Lehrbach, P.R. and Timmis, K.N. (1986b) Vector for regulated expression of cloned genes in a wide range of gram-negative bacteria. *J. Bacteriol.* **167**, 447–54.

Molin, S., Klemm, P., Poulsen, L.K., Biehl, H., Gerdes, K. and Andersson, P. (1987) Conditional suicide system for containment of bacteria and plasmids. *Bio/Technology* **5**, 1315–18.

Neidhardt, F.C., Vaughn, V., Phillips, T.A. and Bloch, P.L. (1983) Gene-protein index of *Escherichia coli* K-12. *Microbiol. Rev.* **47**, 231–84.

O'Farrell, P.H. (1975). High resolution two-dimensional electrophoresis of proteins. *J. Biol. Chem.* **250**, 4007–21.

Palmer, L.M., Baya, A.M., Grimes, D.J. and Colwell, R.R. (1984) Molecular, genetic and phenotypic alteration of *Escherichia coli* in natural water microcosms containing toxic chemicals. *FEMS Microbiol. Lett.* **21**, 169–73.

Pritchard, P.H., O'Neill, E.J., Spain, C.M. and Ahearn, D.G. (1987) Physical and biological parameters that determine the fate of *ρ*-chlorophenol in laboratory test systems. *Appl. Environ. Microbiol.* **53**, 1833–8.

Ramos, J.L., Mermod, N. and Timmis, K.N. (1987). Regulatory circuits controlling transcription of TOL plasmid operon encoding meta-cleavage pathway for degradation of alkylbenzoates by *Pseudomonas. Molec. Microbiol.* **1**, 293–300.

Rojo, F., Pieper, D.H., Engesser, K.-H., Knackmuss, H.-J. and Timmis, K.N. (1987) Assemblage of *ortho* cleavage route for simultaneous degradation of chloro- and methylaromatics. *Science* **238**, 1395–8.

Sagik, B.P., Sorbert, C.A. and Moore, B.E. (1981). The survival of EK1 and EK2 systems in sewage treatment plant models. In Levy, S.B., Clowes, R.C. and Koenig, E.L. (Eds.) *Molecular Biology, Pathogenicity, and Ecology of Bacterial Plasmids*, Plenum Press, New York, pp. 449–60.

Saye, D.J., Orgunseitan, O., Sayler, G.S. and Miller, R.V. (1987) Potential for transduction of plasmids in a natural freshwater environment: effect of plasmid donor concentration and a natural microbial community on transduction in *Pseudomonas aeruginosa. Appl. Environ. Microbiol.* **53**, 987–95.

Sayler, G.S., Shields, M.S., Tedford, E.T., Breen, A., Hooper, S.W., Sirotkin, K.M. and Davis, J.W. (1985) Application of DNA–DNA colony hybridization to the detection of catabolic genotypes in environmental samples. *Appl. Environ. Microbiol.* **49**, 1295–303.

Schwartz, D.C. and Cantor, C.R. (1984) Separation of yeast chromosome-sized DNAs by pulsed field gradient gel electrophoresis. *Cell* **37**, 67–75.

Simon, R., Priefer, U. and Pühler, A. (1983) A broad host range mobilization system for *in vivo* genetic engineering: transposon mutagenesis in gram negative bacteria. *Bio/Technology* **1**, 784–91.

Smith, G.A., Nickels, J.S., Kerger, B.D., Davis, J.D., Collins, S.P., Wilson, J.T., McNabb, J.F. and White, D.C. (1986) Quantitative characterization of microbial biomass and community structure in subsurface material: a prokaryotic consortium responsive to organic contamination. *Can. J. Microbiol.* **32**, 104–11.

Southern, E. (1975) Detection of specific sequences among DNA fragments separated by gel electropheresis. *J. Molec. Biol.* **98**, 503–17.

Stoker, N.G., Fairweather, N.F. and Spratt B.G. (1982). Versatile low-copy-number plasmid vectors for cloning in *Escherichia coli. Gene* **18**, 335–41.

Stotzky, G. and Babich, H. (1984) Fate of genetically-engineered microbes in natural environments. *Recombinant DNA Technical Bull.* **7**, 163–88.

Stotzky, G. and Krasovsky, V.N. (1981) Ecological factors that affect survival, establishment, growth and genetic recombination of microbes in natural habitats. In: Levy, S.B., Clowes, R.C. and Koenig, E.L. (Eds.) *Molecular Biology, Pathogenicity, and Ecology of Bacterial Plasmids*, Plenum Press, New York, pp. 31–42.

Tait, R.C., Close, T.J., Lundquist, R.C., Hagiya, M., Rodriguez, R.L. and Kado, C.J. (1983) Construction and characterization of a versatile broad-host range DNA cloning system for gram-negative bacteria. *Bio/Technology* **1**, 269–75.

Timmis, K.N. and Manning, P.A. (1986) Virulence factors in enteropathogenic bacteria. In: Falmagne, P., Alouf, J.E., Fehrenbach, F.J., Jeljaszewicz, J. and Thelestam, M. (Eds.) *Bacterial Protein Toxins*, Gustav Fischer Verlag, Stuttgart, pp. 301–13.

Timmis, K. and Winkler, U. (1973) Isolation of covalently-closed circular deoxyribonucleic acid from bacteria which produce exocellular nuclease. *J. Bacteriol.* **113**, 508–9.

Towbin, H., Staehelin, T. and Gordon, J. (1979) Electrophoretic transfer of proteins from polyacrylamide gels to nitrocellulose sheets: procedure and some applications. *Proc. Natl Acad. Sci. U.S.A.* **76**, 4350–4.

Weinberg, S.R. and Stotzky, G. (1972) Conjugation and genetic recombination of *Escherichia coli* in soil. *Soil Biol. Biochem.* **4**, 171–80.

Willetts, N. and Wilkins, B. (1984) Processing of plasmid DNA during bacterial conjugation. *Microbiol. Rev.* **48**, 24–41.

10 Patterns of Establishment and Spread of Animal Viruses

FRANK FENNER
The John Curtin School of Medical Research, The Australian National University, Canberra, Australia

10.1 INTRODUCTION

In the mid-1960s my colleagues and I were engaged in experiments on animal virus genetics, using vaccinia virus and a vaccine strain of poliovirus. One of my research students was Joe Sambrook, later to become head of the Tumor Virology Laboratory at Cold Spring Harbor. I asked Sir Macfarlane Burnet to examine his thesis. It was a good thesis and no problems arose with its acceptance, but it prompted Burnet (1966) to write a provocative article in the journal *Lancet* entitled 'Men or molecules? A tilt at molecular biology'. He commented that, although genetic manipulation of bacteriophages was harmless, 'its inevitable extension to such readily manipulated viruses as poliomyelitis and vaccinia raises some very serious dangers'. He went on to envisage the possible consequences of the emergence of a genetic variant of poliovirus with high virulence but serologically quite different from polioviruses 1, 2, and 3. If such a virus was produced, he foresaw 'the almost unimaginable catastrophe of a "virgin-soil" epidemic of poliomyelitis involving all the populous regions of the world'.

Some years later recombinant-DNA techniques were discovered. Some of the molecular biologists who developed these methods shared Burnet's misgivings, and convened the Asilomar Conference. Their concern was largely with the potential danger to humans of the products of recombinant-DNA techniques utilizing tumor viruses like SV40 and the retroviruses. There were many biochemists at the Asilomar Conference but few epidemiologists, and the result was something of an overreaction, because none of the prominent speakers knew much about how infectious agents spread between vertebrate hosts. We have come a long way since then, and this conference is an attempt to address the question, can organisms modified by recombinant-DNA techniques become established and spread in the environment. and what will be the consequences if they do, to man and his domesticated animals and plants and to ecosystems? Are the consequences

Introduction of Genetically Modified Organisms into the Environment. Edited by Harold A. Mooney and Giorgio Bernardi

likely to be more or less advantageous and more or less harmful to man and to ecosystems than the introduction of organisms changed by other methods, such as conventional plant and animal breeding?

Coming to the topic assigned to me, 'Patterns of establishment and spread', I will restrict my talk to consideration of the question: how are animal viruses transmitted from one animal to another and how do they become established and survive in the environment, i.e. in populations of animals? Most of the general principles that I will outline are equally applicable to other infectious agents, such as bacteria and protozoa. After briefly considering some aspects of epidemiological theory (for reviews, see Evans, 1989; Fenner *et al.*, 1987; White and Fenner, 1986), I will look at some examples with 'normal' viruses that may be taken as models of the possible establishment of genetically modified organisms in the environment. I will then discuss live virus vaccines, for most of these are produced by the genetic manipulation of virulent viruses, and finally describe recombinant vaccinia virus vaccines, which are currently being considered for use in human and veterinary medicine, and thus for widespread and large-scale introduction into the environment.

10.2 VERTICAL TRANSMISSION

To become established in the environment, viruses must be transmitted from one animal to another, in an ongoing series of infections. They may be transmitted from one animal to another either vertically or horizontally. 'Vertical transmission' means transmission from parents to offspring via the egg or sperm, either as a provirus integrated in the cell's DNA or in mammals, via the placenta or during the birth process. Other modes of transmission are designated 'horizontal'. When considering the establishment of any novel virus in the environment, the initial problem is of course horizontal transmission, but if this occurred, vertical transmission could be important in helping to ensure the indefinite maintenance in a vertebrate population of a virus that had been introduced by horizontal transmission.

10.3 HORIZONTAL TRANSMISSION

Horizontal transmission can occur across any surface of the body: skin, respiratory mucous membrane, genital tract, and, more rarely, the urinary tract or conjunctiva. The route of infection is fairly characteristic of the viral group (Table 10.1), although some viruses are transmitted by more than one route.

Table 10.1 Some examples of routes of infection of viruses of different genera

Route	Virus and disease			
	Localized		Generalized	
	Genus	Disease	Genus	Disease
Respiratory	*Rhinovirus*	Common cold	*Orthopoxvirus*	Smallpox
	Orthomyxovirus	Influenza	*Morbillivirus*	Measles
	Adenovirus (most)	Pharingitis	*Herpesvirus*	Chickenpox
Alimentary (per os)	*Enterovirus* (most)	Nil	*Enterovirus* (few)	Poliomyelitis
	Adenovirus (few)	Nil		Hepatitis A
	Rotavirus	Diarrhea		
(per rectum)			*Lentivirus* (HIV)	AIDS
Skin	*Papillomavirus*	Warts		
	Poxviridae	Molluscum contagiosum		
Skin (injection)	*Orthopoxvirus*	Vaccinia	Arboviruses	Yellow fever
				Encephalitis
Congenital or perinatal			*Lentivirus* (HIV)	AIDS
			Rubella virus	Congenital defects
			Hepadnavirus	Hepatitis B carcinoma
Genital tract	*Papillomavirus*	Genital warts	Herpes simplex	Genital herpes
			Lentivirus (HIV)	AIDS

10.4 SPREAD IN THE BODY

Whatever their route of entry, viruses may remain localized, for example in the skin or upper or lower respiratory tract, or they may spread to other parts of the body, usually via the lymphatic system and the blood stream. Viruses introduced by inoculation, either by arthropods or by needle, almost always produce generalized infections. Whether a virus spreads within the animal or not affects the nature of the disease it causes and how it spreads to other animals. Most respiratory viruses multiply only in the respiratory tract and are excreted by sneezing, coughing, etc. A few, such as measles and chickenpox, enter via the respiratory tract and produce a generalized disease, but spread to other animals occurs from lesions produced in the upper respiratory tract during the rash stage of the disease, 14–21 days after infection. Likewise, most viruses that replicate in the alimentary tract remain localized, but a few, like polioviruses and hepatitis A virus, can cause generalized infections.

10.5 SHEDDING

To survive in nature, infectious agents must either be transmitted vertically, in the germplasm or during the birth process, or they must be shed, so that they can infect other individuals by horizontal infection. Shedding usually occurs via the same surface as the virus enters, but it may occur for only a brief period, as in common colds, for several weeks, as in many enterovirus infections, or recurrently, perhaps throughout life, as in herpesvirus and lentivirus infections.

10.6 INAPPARENT INFECTIONS

In closing this introductory section, we need to note the important distinction to be made between 'infection' and 'disease'. Survival of a virus in nature depends upon the maintenance of infection; the occurrence of disease is neither required nor necessarily advantageous. Infection without disease is called 'inapparent' or 'subclinical' infection.

Overall, inapparent infections are much more common than those that result in disease; for example in poliomyelitis due to wild-type polioviruses there may be hundreds of inapparent infections for every paralytic case, in arbovirus encephalitis the ratio of inapparent infections to encephalitis is 1000 to 1. Although clinical cases may be somewhat more productive sources of infectious virus than inapparent infections, the latter are usually more numerous and do not restrict the movement of infected individuals, and thus they provide an important method of viral dissemination.

10.7 SURVIVAL OF VIRUSES IN NATURE

I will now review some of the factors that affect the survival of viruses in nature (Table 10.2).

10.7.1 TRANSMISSIBILITY AND VIRULENCE

The best-documented demonstration of how the virulence of a virus directly affects the probability of its survival in nature is myxomatosis of rabbits. In this disease the virus is mechanically transmitted by biting arthropods ('flying pins'), and transmission is most effective when the diseased rabbit maintains highly infectious skin lesions for a long period. Very virulent strains of myxoma virus kill rabbits too quickly for optimum transmission, whereas the lesions caused by highly attenuated strains have too low a concentration of virus particles in the skin lesions for the mouthparts of biting arthropods to become contaminated (Fenner and Ratcliffe, 1965). On theoretical grounds, we predicted that viruses at either extreme of the virulence range would not survive in nature, and this has been confirmed by data from Australia and Europe over the last 35 years (Fenner and Ross, 1990).

10.7.2 CRITICAL COMMUNITY SIZE

The importance of this factor is best exemplified by two common infections of children, measles and chickenpox, which can survive indefinitely in populations of very different sizes, because of different patterns of shedding (Black, 1980). The incubation period of measles is about 12 days and an attack is followed by solid immunity to reinfection. Once a person has had measles and been infectious for about a week at the height of the disease,

Table 10.2 Factors affecting the survival of viruses in nature

Factor	Example
Transmissibility associated with moderate virulence	Myxoma virus (rabbits)
Large populations	Measles virus
Persistent or recurrent infectivity	Herpesviruses Hepatitis B virus
Transmission in germplasm	Retroviruses (birds and rodents)
Antigenic drift and shift	Influenza A virus
Animal reservoir	
With arthropod transmission	Arboviruses
Without arthropod transmission	Arenaviruses

he or she never becomes infectious again, i.e. recurrent infectivity is unknown. Persistence of measles virus in a community therefore depends upon a continuous supply of susceptible subjects, usually supplied by new births. Measles spreads more readily in winter, and the infection of many susceptibles at this time may cause difficulty in summer, when its transmission is less efficient. In urban areas, where it spreads quickly, as many as 4000 births a year may be needed to sustain the virus (Bartlett, 1960). In rural areas, where it spreads more slowly, a smaller number may suffice, but 1000 births per year seems to be the lower limit (Black, 1966). These numbers of births imply populations of between 200 000 and one million persons. Large epidemics can occur in smaller populations and may persist for a considerable time, but sooner or later 'fade-out' occurs and the disease disappears until there is a new introduction.

Chickenpox is also an acute exanthem in which infection is followed by lifelong immunity to reinfection, but it differs radically from measles in the pattern of shedding of the virus. Outbreaks of chickenpox are due to respiratory spread to contacts during the acute stage of the disease, but as in all herpesvirus infections, recurrent shedding occurs, often many years after the initial attack. Varicella virus, after being latent in the sensory nerve ganglia, may be reactivated and cause a disease called zoster (shingles). For this reason, the virus is usually called 'varicella-zoster virus'. Although zoster is only about one quarter as infectious as chickenpox (secondary attack rates of 15%, compared with 60–70% for varicella), it can produce chickenpox in susceptible children (Hope-Simpson, 1952, 1954; Gordon, 1962). Since zoster may occur 50 or more years after an attack of chickenpox, varicella-zoster virus requires a dramatically smaller critical community size for indefinite persistence; less than 1000, compared with about 500 000 for measles.

10.7.3 PERSISTENT INFECTIONS

Viral persistence is of considerable epidemiological importance because it is usually, although not always, associated with continuous or recurrent viral shedding, a mechanism that greatly enhances the likelihood of viral survival by providing recurring opportunities for horizontal spread to other host animals. Many viruses characteristically cause persistent infections. Herpesviruses are the best studied; lentiviruses, including the human immunodeficiency viruses, constitute another important group in which virus persists for the life of the host.

10.7.4 ANTIGENIC DRIFT AND SHIFT

The influenza viruses provide the best example of how changes in the antigens involved in evoking a protective immune response may enable a virus to survive in human populations (Murphy and Webster, 1985). Between pandemics, influenza survives in man by sequential mutations in critical antigenic sites in the haemagglutinin which allow the virus to spread in a population in which many individuals have antibodies due to prior infection with an earlier version of the same viral subtype. Such changes in the antigenic structure of the haemagglutinin are called 'antigenic drift'. Sometimes reassortants arise by recombination between a human and an animal (swine or avian) influenza A virus which result in the introduction of a novel (for man) haemagglutinin or neuraminadase gene. Sometimes such a reassortant spreads in humans, and since the global human population is non-immune, a pandemic of influenza results. This process is called 'antigenic shift'; two instances have been documented since the human influenza virus was first isolated in 1933: Asian influenza in 1957 (H1N1 to H2N2) and Hong Kong influenza in 1968 (H2N2 to H3N2).

10.7.5 MULTIPLE-HOST INFECTIONS

Most viral infections of man and domestic animals are maintained in nature within human or domestic animal populations. However, a number of viruses can spread naturally between different species of vertebrate host; the term 'zoonosis' is used to describe a multiple-host infection that is naturally transmissible from lower animals to man. Viral infections of livestock may also have wildlife reservoirs. Many such diseases are caused by arboviruses, and can be spread from one species of vertebrate host to another, as well as from one member to another of the same species, by the bite of an infected arthropod.

When arthropods are active, arboviruses multiply alternately in vertebrate and invertebrate hosts. It is more difficult to understand what happens to the virus during the winter months in temperate climates, when the arthropod vectors are frequently inactive. An important mechanism of such 'overwintering' is transovarial transmission from one generation of arthropods to the next. Long known to occur with the tick-borne flaviviruses, this type of vertical transmission has more recently been discovered to occur also with some mosquito-borne bunyaviruses and flaviviruses as well.

10.8 SOME EXAMPLES OF ESTABLISHMENT AND SPREAD IN NEW ENVIRONMENTS

So much for background epidemiology. I will now examine a number of situations which illustrate various facets of the establishment of different viruses in geographically or ecologically novel environments. As is the case with biological invasions of plants and animals, my examples involve viruses that had previously established ecological niches for themselves in other environments; genetically modified viruses may be incapable of spreading in nature under any ecological conditions. Indeed that has been a condition for the licensing of some viral vaccines, e.g., rubella virus vaccine. I will begin with some examples of poxvirus infections.

10.8.1 MYXOMA VIRUS

Experiments in Californian lagomorphs

Leporipoxvirus, the genus to which myxoma virus belongs, evolved in parallel with lagomorphs and squirrels in North America, with later extension to South America. Myxoma virus produces a trivial lump in the skin of its natural reservoir hosts, *Sylvilagus bachmani* in California and *S. brasiliensis* in Central and South America. The level of adaptation of myxoma viruses for survival in their particular natural hosts is illustrated by comparisons of the behaviour of the Californian strain of myxoma virus in different species of Californian rabbits (Table 10.3) and of different strains of myxoma virus in the natural host of the Californian strain of virus, *S. bachmani* (Table 10.4).

Table 10.3 shows that the natural reservoir host, *S. bachmani*, is an effective source of infection for mosquitoes, which produce lesions when they bite any of several species of recipient host. However, when transfer experiments were attempted from the lesions of the recipient hosts, only those mosquitoes that probed through the lesions in *S. bachmani* produced infection in the very susceptible indicator host, *Oryctolagus cuniculus*. Experiments with the Californian strain of myxoma virus, a virulent Brazilian strain (Lausanne), and a somewhat attenuated derivative of a Brazilian

Table 10.3 Failure to transmit Californian myxoma virus by mosquito bite from lesions on all *Sylvilagus* species except *S. bachmani*[a]

Donor host	Mosquito transfer	Recipient host	Second mosquito transfer	Challenge host
S. bachmani	26/54[b,d]	*S. bachmani*	153/253[c,d]	*O. cuniculus*
S. bachmani	16/20	*S. audubonii*	0/35	*O. cuniculus*
S. bachmani	7/12	*S. nuttallii*	0/18	*O. cuniculus*
S. bachmani	3/3	*S. idahoensis*	0/15	*O. cuniculus*
S. bachmani	4/7	*S. floridanus*	0/21	*O. cuniculus*

[a]Based on Regnery and Marshall (1971).
[b]Numerator, positive results on attempted transfers from *S. bachmani* to recipient host indicated; denominator, number of transfers attempted.
[c]Numerator, positive results on attempted transfer from recipient host to *O. cuniculus*; denominator, numbers of transfers attempted.
[d]The rather low efficiency of transfers from *S. bachmani* (first line) is due to the inclusion of many attempts at unfavourable times, e.g. with poorly developed or very old tumors. All other transfer attempts (lines 2–5) were made at the optimum times.

Table 10.4 Failure to transfer viruses other than Californian myxoma virus from *S. bachmani* by means of mosquito bites[a]

	Virus strain		
	Attenuated Australian (KM13)	Virulent South American (Lausanne)	Virulent Californian
Ratio			
Titre in *O. cuniculus* to titre in *S. bachmani*	1000	1	1
Mosquito bite transfers from lesions in *S. bachmani*	...[b]	0/20	16/20
Virus titre[c]; skin slices from *S. bachmani*	...	10^4	10^8

[a]Modified from Marshall and Regnery (1963).
[b]... indicates not done.
[c]Infectious doses per gram of tissue.

strain that had occurred naturally in Australia (KM13) showed that only the Californian strain of myxoma virus produced lesions in *S. bachmani* from which mosquitoes could acquire enough virus particles to transmit infection to the indicator host, *O. cuniculus* (Table 10.4).

Infection of *Oryctolagus cuniculus*

By an accident of nature, myxoma virus produces a devastating generalized infection in the European rabbit, *Oryctolagus cuniculus*, with a case fatality rate of almost 100%. This property led to its use for the biological control of rabbits in Australia and Europe (Fenner and Ratcliffe, 1965). Here I will consider only how myxoma virus became established and spread in these new environments.

Experiments in dry parts of Australia in the 1940s showed that if Australian wild rabbits were infested with the fleas of certain marsupials (*Echidnophaga myrmecobii*) infection spread within but not between warrens. Results were little better in the first trials undertaken after the Second World War, when virus was introduced into four different sites in the Murray Valley Region in Victoria over the period August–November 1950. In the best documented trial, an introduction in October 1950 went through four generations of transmission, each causing progressively fewer infections. In spite of a number of deaths, the rabbit population rose from 700 to over 1000 and the infection died out (Myers, 1954).

Establishment and spread in the Australian wild rabbit population occurred a few weeks after the conclusion of these unsatisfactory trials in the Murray Valley, when the disease spread over a vast area of southeastern Australia because of the activity of very large populations of rabbit-biting mosquitoes that regarded rabbits as a highly preferred source of blood meals. They picked up virus particles as they probed through skin lesions, and then acted as 'flying pins' to transfer them to other rabbits. Earlier failures to establish the disease in Australian wild rabbits were due to faulty choice of location or season for the attempted introduction.

The establishment of myxomatosis in Europe followed quite a different pattern. Hearing of the devastating outbreaks of myxomatosis in Australia, a French pediatrician whose estate was plagued with rabbits obtained a strain of the virus from a friend in Switzerland and on 14 June 1952 inoculated two wild rabbits with the virus. By the end of August all rabbits on his estate were dead and outbreaks had occurred up to 45 km away—myxomatosis was established in Europe. Thatcherites would say that this success was inevitable, because private enterprise rather than a government authority was involved, but in fact the French doctor had by chance chosen the optimum time to ensure that mosquito transmission would occur. The contrast between the experiences with myxomatosis in Australia and France illustrates that there is a large element of chance in the process of the establishment and spread of an infectious agent in a new environment.

There are still mysteries about how myxoma virus survives through the Australian winter, when mosquitoes are inactive. Myxoma virus probably did and does die out in many localities, but we believe that it persists in some places because of low-level mosquito transmission and the presence of a few rabbits that remain infectious for a long period. The difficulty of survival over the winter period was a powerful selective force for the emergence and perpetuation of viruses that were somewhat attenuated (see above). Because of the mobility of the mosquito vector, myxomatosis can spread over wide areas each summer, if conditions are suitable for the buildup of large mosquito populations. In Britain, and in parts of Australia where they have become established since their introduction in the 1970s, rabbit fleas (*Spilopsyllus cuniculi*) provide a vector that operates the year round.

10.8.2 COWPOX VIRUS

Cowpox virus, which is thought to be the virus used by Jenner in 1796 to protect children against smallpox, was for a long time recognized only as a disease of cows and persons who milked cows. However, it was and is a rare disease in both cows and humans, and we now know that it also affects domestic cats and a variety of zoo animals (Fenner *et al.*, 1989). It is quite clear that it does not survive in nature as a disease of any of these hosts; since recurrent infectivity does not occur, infection is followed by lifelong immunity, and all of these animals have slow population turnover times. Cowpox virus has been recovered from wild rodents in Turkmenia and from rat colonies in Moscow and Poland. We now suspect that cowpox virus survives in wild rodents, which have a large population size and rapid turnover, and infection occasionally spreads from these reservoir hosts to the larger animals.

10.8.3 HUMAN MONKEYPOX

In 1970, about eight months after the last case of smallpox in an area of tropical rain forest in Zaire, a case of vesiculopustular disease clinically indistinguishable from smallpox was reported; subsequently other sporadic cases were recognized in several countries in western Africa. Smallpox did not occur as a 'sporadic' disease; every case derived from contact with another human case. The occurrence of sporadic cases of any such disease implies a wildlife source; this sporadic disease was in fact due to another species of *Orthopoxvirus*—monkeypox virus. Intensive studies in the field and in the laboratory have shown that monkeypox virus probably survives in nature in Zaire as a disease of squirrels, and that occasionally chimpanzees, a number of species of monkeys, and man acquire the infection (Jezek and Fenner, 1988). In the context of this conference, the interesting feature of monkeypox is that although human-to-human infection does occur (Table 10.5), it is so inefficient that, unlike smallpox, monkeypox cannot survive as a human infection—it remains a zoonosis that occasionally gives rise to secondary human cases.

Table 10.5 Human monkeypox: occurrence of primary and secondary cases in Zaire, 1982–84

Year	Primary cases[a]		Person-to-person infection		Total
	Isolated primary case	Presumed co-primary case	Presumed secondary case	Presumed tertiary or later case	
1982	20	4	13	3	40
1983	37	21	19	7	84
1984	43	19	17	7	86
Total	100 (47.6%)	44 (21.0%)	49 (23.3%)	17 (8.1%)	210

[a]Presumably infected from an animal source.

10.9 LIVE VIRUS VACCINES

Other than Jenner's smallpox vaccine and a few later parallels (turkey herpesvirus for Marek's disease; fibroma virus for myxomatosis) all live virus vaccines have been genetically modified, in that they are attenuated variants of the virulent pathogen concerned. Among those used for human vaccination, the first was Pasteur's 'fixed' strain of rabies virus, followed many years later by yellow fever vaccine, then the poliovaccines and later by measles, rubella, and mumps virus vaccines.

With all of these viruses, the element of possible spread from the vaccinated host arose. Yellow fever vaccine does not produce sufficient viraemia to infect mosquitoes, and neither measles nor mumps vaccines, administered by injection or, for measles, by aerosol, are shed from the oropharynx in quantities adequate to infect contacts. This aspect of viral behaviour was of particular importance in designing rubella virus vaccine (Scott and Byrne, 1971), for the objective in this instance was to protect the fetus and one did not wish by accident to infect a pregnant woman who was in contact with a vaccinated child.

10.9.1 POLIOVIRUS VACCINES

From the point of view of this conference, the Sabin oral poliovirus vaccines are of particular interest. Wild-type polioviruses cannot survive in small communities, although their critical community size is smaller than that of measles, because shedding may persist for several weeks. However, they die out in small isolated groups like the Eskimos, but if a strain that causes paralysis is introduced into such a 'virgin-soil' community it may cause a catastrophic outbreak.

In the late 1950s attenuated strains of each of the three serotypes of poliovirus had been developed by Albert Sabin by tissue culture passage and selection for loss of neurovirulence for monkeys. In 1961–2 they were licensed in the United States, and have proved to be spectacularly successful in controlling poliomyelitis. The question I wish to explore today is how well these vaccine viruses, administered each year to millions of children, have survived in nature.

When we talk about the polioviruses we are discussing three distinct viruses, both for wild-type and vaccine strains. I will concentrate my remarks on poliovirus type 1 and the corresponding vaccine strain, because more detailed information is now available about these as a result of recent sequencing studies of a large number of isolates (Rico-Hesse *et al.*, 1987).

Poliovirus type 1

In selecting for vaccine strains, Sabin was concerned with two factors:

absence of neurovirulence and capacity to immunize after feeding. In both respects the Sabin type 1 virus is an excellent vaccine; it has eliminated endemic wild-type 1 from the United States (Kim-Farley *et al.*, 1984) and it caused only one vaccine-associated case for every 7 to 19 million first doses of vaccine administered (Nkowane *et al.*, 1987) (the range in possible frequency arises from the fact that more than one serotype of vaccine virus was recovered from 9 out of the 14 cases that yielded type 1 vaccine virus). I have already made one comment of interest for this audience, namely that type 1 poliovaccine eliminated wild-type 1 poliovirus from the United States. A similar result has been reported from the Scandinavian countries, where inactivated poliovirus vaccines have eliminated endemic wild polioviruses. This was an immunological exclusion, not one based on interference or some kind of intraspecific competition of the vaccine and wild-type strains.

We also have some data on the survival of poliovirus vaccines in communities in various parts of the world. Some of this comes from vaccine-associated cases, which always occur in vaccinated individuals or their close contacts; vaccine strains have never been known to cause an epidemic. The other data come from a variety of sources. As I said earlier, Sabin selected for absence of virulence and ability to immunize the inoculated individual, not on the ability of the virus to survive in nature and to spread from one person to another. These vaccine strains in fact do not survive well in the environment, for several reasons:

1. At least in cultured cells, the yield is far below that of wild viruses; for type 1, the ratio is about 1:10 000.
2. The vaccine strains are physically more fragile than wild-type strains and are more easily degraded by heat and other environmental insults.
3. They shed for short periods only—for type 1 for only 2 weeks, compared with 12–16 weeks for wild type.

In the developed countries using oral poliovirus vaccines, the only polioviruses now isolated from sewage are vaccine derived. This is due to the fact that in such countries enterovirus infection is seasonal (summer–autumn), and most new susceptibles are vaccinated before the season begins. Since vaccination is given the year round, vaccine viruses are constantly being excreted and are regularly found in sewage. Analysis of these excreted viruses shows that they have all undergone mutational changes and quite a number are intratypic recombinants (Rico-Hesse *et al.*, 1987). However, none of them survives in nature anything like as well as wild polioviruses.

An interesting facet of the spread of infectious agents around the world in these days of extensive air travel is that of the 30 sporadic cases and three epidemics of paralytic poliomyelitis in the United States due to poliovirus type 1 over the period 1974–84, all were due to imported viruses, most of these tested coming from Mexico, Central America, and the Middle

East. A well-studied outbreak that moved in an unvaccinated religious community from the Netherlands to Canada to the United States in 1978–9 apparently originated in Turkey (Rico-Hesse *et al.*, 1987).

10.9.2 RECOMBINANT VACCINIA VIRUS VACCINES

There is a great deal of interest these days in the production of new types of vaccine. Three main stratagems are being explored: antigens produced by recombinant-DNA methods in *E. coli*, yeasts, or animal cells, synthetic peptides that are identical to critical antigenic sites, and vaccines constructed by introducing relevant genes into a vector virus or bacterium, which is then administered as a live vaccine. I will discuss only the latter, and of the variety of vectors available, only vaccinia virus, the agent used for smallpox vaccination (for reviews see Fenner *et al.*, 1989; Moss and Flexner, 1987; Mackett and Smith, 1986).

Vaccinia virus, which multiplies in the cytoplasm of the cell, has its own promoters. Recombinant vaccinia viruses can be constructed according to the protocol illustrated in Figure 10.1, by simultaneous vaccinia virus infection and transfection of cells with a plasmid containing the relevant foreign gene, a vaccinia promoter, and flanking fragments of the vaccinia virus thymidine kinase (TK) gene. Homologous recombination incorporates the foreign gene into the vaccinia virus genome and the recombinant viruses are selected for their TK^- character.

A large number of genes from a variety of viruses and protozoa have been incorporated into such vectors. They are expressed when cells are infected and they immunize experimental animals. As used in the smallpox eradication programme, smallpox vaccine had great advantages over all other human vaccines used in tropical developing countries. It was extremely heat stable, it could easily be produced, even in developing countries, and it could be administered by a bifurcated needle or a jet injector without the need for a syringe (Fenner *et al.*, 1988). Recombinant vaccinia virus vaccines have all these advantages. The disadvantage of smallpox vaccine was that occasionally it caused severe complications, at rates of 3–10 per million primary vaccinations. Whether this is acceptable depends upon the risk against which protection is sought—it was certainly worth taking when and where smallpox had a case-fatality rate of 25%.

With our increasing knowledge of the molecular biology of vaccinia virus, it is possible to improve the performance of such vaccines by three types of stratagem: lowering the virulence of the virus, increasing the expression of the inserted gene, and possibly by increasing the immune response in the vaccinated subject. Considerable improvements have been obtained by the use of some of these approaches. The use of the vaccinia virus TK gene for insertion leads to a dramatic reduction in the virulence of the virus and hence its propensity to cause complications (Buller *et al.*, 1985). Indeed,

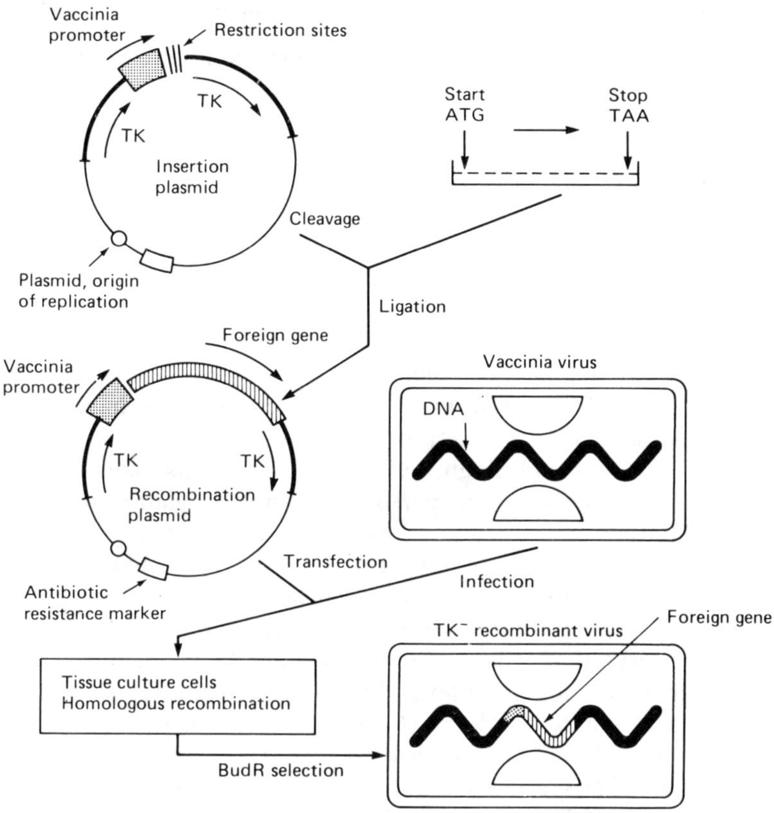

Figure 10.1 Method of constructing a recombinant vaccinia virus carrying a selected gene from another source (virus, protozoan, or eukaryote). TK = thymidine kinase gene of vaccinia virus; BudR = bromodeoxyuridine. (Courtesy Dr B. Moss.)

there is a risk that TK⁻ mutants may be overattenuated and not multiply enough to produce a good immune response. This risk may be partially counteracted by greatly increasing the expression of the inserted gene. One way of doing this is by the production of a vaccinia virus recombinant by infection of cells with two recombinants, one with the foreign gene flanked by the T7 bacteriophage polymerase promoter and another containing the T7 polymerase gene (Fuerst *et al.*, 1987). This has resulted in a tenfold increase in expression of the inserted gene. Another possibility, still being explored, is that the insertion of genes for lymphokines may dramatically augment the host's immune response. So far only interleukin-2 has been studied, but thymus-deficient 'nude' mice and heavily irradiated mice infected with vaccinia containing interleukin-2 were protected against a million times the usual lethal does of vaccinia virus (Ramshaw *et al.*, 1987).

Establishment of vaccinia virus in the environment

Efforts will soon be made to use such vaccines, so that consideration needs to be given to their potential for spread. During the period when smallpox vaccination was widely practiced, contact infection of vaccinated persons occasionally occurred, and could be serious if eczematous children were affected (Fenner *et al.*, 1988). The occurrence in cows of 'cowpox' due to vaccinia virus acquired from vaccinated humans has been described on a number of occasions, but only one outbreak of vaccinia in man has been reported, in association with a self-limited outbreak of vaccinia virus-induced 'cowpox' (Lum *et al.*, 1967).

The only known instance of the probable escape of vaccinia virus into the environment and its maintenance in nature over a period of years relates to buffalopox in India. For many years, veterinarians have described a disease that they called buffalopox among water buffalos in India, Egypt, and the USSR (Lal and Singh, 1977). Most outbreaks could be traced to the transfer of infection from persons vaccinated against smallpox to the buffalos that they milked, and transfer between buffalos was effected by milkers, or sometimes by sucking calves. However, smallpox vaccination was discontinued world-wide in 1980, yet as recently as 1986 outbreaks of buffalopox have been reported in Maharashtra State in India. All strains of 'buffalopox' virus that have been studied have been identified as vaccinia virus, but their restriction patterns differ slightly from those of strains used for smallpox vaccination in India (M. Richardson and K. Dumbell, personal communication, 1987). The implication is that vaccinia virus is surviving in nature in some places in India and causing recurrent outbreaks of buffalopox. It is said that buffalopox is a much more chronic infection than cowpox, but I find it hard to believe that it is surviving in buffalos, perhaps with humans as intermediate hosts. The problem merits proper epidemiological investigation.

Even with this precedent, I believe that the new recombinant vaccinia virus vaccines should be safe, because they are much less virulent than the strains used for smallpox vaccine and are correspondingly less likely to spread from one animal to another.

10.10 CONCLUSIONS

Experience with the introduction of animal viruses into geographically and ecologically novel environments confirms experience with biological invasions of plants and animals, namely that the risk of establishment and spread under such conditions is unpredictable; chance factors, including timing, are critical. Although insect viruses are used for the control of some insect pests, deliberate large-scale introductions of viruses of vertebrates in situations in which they may be released into the environment will occur most frequently in association with immunization against infectious diseases—

those of man, of domestic animals, and (with rabies virus) of wildlife. Provided that care is taken in the selection and design of the viruses used, whether they are ordinary live virus vaccines or vectors for genes from other pathogens, I believe that the benefits of these activities should greatly exceed the risks, but each case will have to be examined on its own merits.

REFERENCES

Bartlett, M.S. (1960) The critical community size for measles in the United States. *J.R. Statist. Soc. Ser. A* **123**, 37–44.

Black, F.L. (1966). Measles endemicity in insular populations: critical community size and its evolutionary implication. *Theor. Biol.* **11**, 207–11.

Black, F.L. (1980) Modern isolated pre-agricultural populations as a source of information on prehistoric epidemic patterns. In: Stanley, N.F. and Joske, R.A. (Eds.) *Changing Disease Patterns and Human Behaviour*, Academic Press, London.

Buller, R.M.L., Smith, G.L., Cremer, K., Notkins, A.L. and Moss, B. (1985) Decreased virulence of recombinant vaccinia virus expression vectors is associated with a thymidine kinase-negative phenotype. *Nature (Lond.)* **317**, 813–15.

Burnet, F.M. (1966) Men or molecules? A tilt at molecular biology. *Lancet* **1**, 37–9.

Evans, A.S. (Ed.) (1989) *Viral Infections of Humans. Epidemiology and Control*, 3rd ed., Plenum, New York.

Fenner, F. and Ratcliffe, F.N. (1965) *Myxomatosis*, Cambridge University Press, Cambridge.

Fenner, F. and Ross, J. (1990) Myxomatosis. In: Thompson, H.V. and King, C. (Eds.) *The Rabbit in Britain, France and Australasia: the Ecology of a Successful Colonizer*, Oxford University Press, Oxford (in press).

Fenner, F., Bachmann, P.A., Gibbs, E.P.J., Murphy, F.A., Studdert, M.J. and White, D.O. (1987) *Veterinary Virology*, Academic Press, Orlando.

Fenner, F., Henderson, D.D., Arita, I., Jezek, Z. and Ladnyi, I.D. (1988) *Smallpox and Its Eradication*, World Health Organization, Geneva.

Fenner, F., Dumbell, K.R. and Wittek, R. (1989) *The Orthopoxviruses*, Academic Press, Orlando.

Fuerst, T.R., Earl, P.L. and Moss, B. (1987) Use of a hybrid vaccinia virus-T7 RNA polymerase system for expression of target genes. *Molec. Cell. Biol.* **7**, 2538–44.

Gordon, J.E. (1962) Chickenpox: an epidemiologic review. *Am. J. Med. Sci.* **244**, 362–89.

Hope-Simpson, R.E. (1952) Infectiousness of communicable diseases in households (measles, chickenpox and mumps). *Lancet* **2**, 549–54.

Hope-Simpson, R.E. (1954) Studies on shingles. Is the virus ordinary chickenpox virus? *Lancet* **2**, 1299–302.

Jezek, Z. and Fenner, F. (1988) *Human Monkeypox, Monograph in Virology*, No. 17, Karger, Basel.

Kim-Farley, R.J., Schonberger, L.B., Nkowane, B.M., Kew, O.M., Bart, K.J., Orenstein, W.A., Hinman, A.R., Hatch M.H. and Kaplan, J.E. (1984) Poliomyelitis in the USA: virtual elimination of disease caused by wild virus. *Lancet* **2**, 1315–17.

Lal, S.M. and Singh, I.P. (1977) Buffalopox—a review. *Trop An. Health Prod.* **9**, 107–12.

Lum, G.S., Soriano, F., Trejos, A. and Llerana, J. (1967) Vaccinia epidemic and epizootic in El Salvador. *Am. J. Trop. Med. Hyg.* **16**, 332–8.

Mackett, M. and Smith, G.L. (1986) Vaccinia virus expression vectors. *J. gen. Virol* **67**, 2067–82.

Marshall, I.D. and Regnery, D.C. (1963) Studies in the epidemiology of myxomatosis in California. III. The response of brush rabbits (*Sylvilagus bachmani*) to infection with exotic and enzootic strains of myxoma virus, and the relative infectivity of the tumors for mosquitoes. *Am. J. Hyg.* **77**, 213–19.

Moss, B. and Flexner, C. (1987) Vaccinia virus expression vectors. *A. Rev. Immunol.* **5**, 305–24.

Murphy, B.R. and Webster, R.G. (1985) Influenza viruses. In: Fields, B.N. *et al.* (Eds.) *Virology*, Raven Press, New York, pp. 1179–239.

Myers, K. (1954). Studies in the epidemiology of infectious myxomatosis of rabbits. II. Field experiments, August–November 1950, and the first epizootic of myxomatosis in the riverine plain of south-eastern Australia. *J. Hyg. (Camb.)* **52**, 47–59.

Nkowane, B.M., Wassilak, S.G.F., Orenstein, W.A., Bart, K.J., Schonberger, L.B., Hinman, A.R. and Kew, O.M. (1987) Vaccine-associated paralytic poliomyelitis. United States: 1973 through 1984. *J. Am. Med. Assoc.* **257**, 1335–40.

Ramshaw, I.A., Andrew, M.E., Phillips, S.M., Boyle, D.R. and Coupar, B.E.H. (1987) Recovery of immunodeficient mice from a vaccinia virus/IL-2 recombinant infection. *Nature (Lond.)* **329**, 545–6.

Regnery, D.C. and Marshall, I.D. (1971) Studies in the epidemiology of myxomatosis in California. IV. The susceptibility of six leporid species to Californian myxoma virus and the relative infectivity of their tumors for mosquitoes, *Am. J. Epidemiol.* **94**, 508–13.

Rico-Hesse, R., Pallansch, M.A., Nottay, B.K. and Kew, O.M. (1987) Geographic distribution of wild poliovirus type 1 genotypes. *Virology*, **160**, 311–22.

Scott, H.D. and Byrne, E.B. (1971) A serologic study of susceptible pregnant women exposed to a state-wide rubella (HPV-77 DE5) immunization campaign. *J. Am. Med. Assoc.* **215**, 609–12.

White, D.O. and Fenner, F. (1986) *Medical Virology*, 3rd ed., Academic Press, Orlando.

11 The Release of Genetically Designed Organisms in the Environment: Lessons from the Study of the Ecology of Biological Invasions

HAROLD A. MOONEY
Department of Biological Sciences, Stanford Univeristy, Stanford, California, USA

and

J.A. DRAKE

Department of Zoology and the Graduate Program in Ecology, University of Tennessee, Knoxville, Tennessee, USA

11.1 INTRODUCTION

Recent concern over the introduction of genetically engineered organisms (GEOs) into the environment has stimulated biologists to look for analogues of the invasion process and the consequences of species introductions to the open environment. Clearly, the best model of the mechanics of species invasions are actual invasion events. A project of SCOPE (Scientific Committee on Problems of the Environment) has recently summarized the global extent and effect of biological invasions across numerous taxonomic groups (Mooney and Drake, 1986; Kornberg and Williamson, 1986; MacDonald *et al.*, 1986; Groves and Burdon, 1986; Drake *et al.*, 1988). This comprehensive review of species introductions provides the basis for our analysis of the potential risks of introducing GEOs to the environment. Here we review the results of this program and relate them to the issues surrounding such releases.

We note at the outset that information on the consequences of species invasions may not be totally germane to considerations of the release of GEOs. Most species invasions, in the sense of the SCOPE program, are

Introduction of Genetically Modified Organisms into the Environment. Edited by Harold A. Mooney and Giorgio Bernardi

invasions from one biogeographic region to another, e.g. species movements from the Old World to the New World, or from North America to Hawaii. Most GEOs, however, will be introduced into very specific environments, often back into environments similar to that from which the immediate progenitors originated. This of course assumes we can engineer leashes of one kind or another into such organisms so that they do not spread beyond the confines of the target system. Nevertheless, an invasion is an invasion; hence we believe we are in a position to identify and address many critical aspects and consequences of such releases.

Throughout this essay we briefly discuss the nature of ecological systems, the interconnectivity, dynamics, and vulnerability of which have been partly responsible for the considerable concern voiced by many ecologists and environmentalists over the envisioned releases of GEOs. We would be remiss at this point, however, if we did not offer an important and critical distinction between 'ecologists' and 'environmentalists'. The lack of this distinction by the scientific and political community in general has added confusion and fueled the controversy over the release of GEOs. Environmentalists, by definition, are concerned with various sociopolitical aspects of environmental quality and management. They may or may not be experts in understanding ecological processes and the organization of ecological systems. Ecologists are scientists concerned with the fundamental properties, processes, and components of ecological systems, whether they be individuals, populations, communities, or ecosystems.

The SCOPE program was initiated in 1982 with the objectives of: (1) determining those properties of invading species that contribute to their success as invaders and (2) identifying the characteristics that make certain ecosystems more vulnerable to invasions than others. The focus of the program was on species that invade natural ecosystems rather than species that invade agroecosystems. There has been a considerable effort in studying the former because of the immediate, identifiable economic losses associated with pest invasions. Concern over the issue of invasions into natural ecosystems does not have such a long history. Indeed, the first major synthesis of knowledge concerning such invasions was produced by Charles Elton in 1958.

The SCOPE program was centered around a series of national meetings designed to summarize current knowledge concerning invasions into various continental regions. National programs included Great Britain (Kornberg and Williamson, 1986), Australia (Groves and Burdon, 1986), the United States (Mooney and Drake, 1986), the Netherlands (Joenje *et al.*, 1987), and South Africa (Macdonald *et al.*, 1986). In addition, a survey of invasions into nature reserves (Usher *et al.*, 1988) and an evaluation of the utility of mathematical models of invasion processes (Drake and Williamson, 1986) were conducted. Finally, a global synthesis summarized these regional assessments (Drake *et al.*, 1988).

11.2 PROGRAM RESULTS

The exchange of organisms between various regions of the biosphere is naturally impeded by obvious physical barriers to migration and species dispersal. These barriers include the proximity and relative global position of continental land masses, mountain ranges, bodies of water, and climatic changes. Prior to humans becoming cosmopolitan in distribution, areas geographically separated shared few species in common. Arguably, the greatest source of taxonomic affinity between the continental land masses probably resulted from species distributions before the break-up of Gondwanaland. This is not to say that intercontinental exchanges did not occur, but just that they were relatively rare events dependent on chance long-distance dispersal by some vector such as migrating birds. There is also evidence of the mixing of biota between continents that had been separated by a water barrier and were joined by a land bridge (Stehli and Webb, 1985).

We are now witnessing a phenomena similar to the creation of land bridges, but on a global scale and at a greatly accelerated rate. These new 'bridges' are humans, and they connect essentially every land mass on the Earth. Humans have purposefully, as well as accidently, introduced organisms into many new environments. Further, they have altered the environments of large areas of the globe, creating new types of habitats and disturbing others. These modifications have undoubtedly changed how vulnerable or resistant these environments are to invasion. For the most part, vulnerability to invasion has increased as a result of human activity. As a result we are seeing a homogenization of the world's biota. Certain species, which in prehistoric times were limited to small isolated regions of the globe, are now widespread.

11.3 THE NUMBERS OF INVADERS

The SCOPE program summarized a great number of examples of invasion rates of organisms into various regions of the world. In many regions of the New World these rates are very high and do not yet show an asymptote. For example, the rate of successful invasion of new species of insects and arthropods into the United States is about 11 species per year (Sailer, 1983). About five species of higher plants successfully colonize eastern Australia each year (Groves, 1986). Although good data do not yet exist on the proportion of invading species which are successful, these values clearly represent a small fraction of the species that annually reach these areas.

Some ecological systems, such as islands, have been considerably more vulnerable to invasion then continental land masses. The Hawaiian Islands are a case in point. Currently 40% of both the birds and higher plants now

Table 11.1 Total number of species and the percentage of established non-native species of the Hawaiian Islands. (From Vitousek *et al.*, 1987)

Group	Total species	Percentage non-native
Invaders		
Flowering plants	~1990	40
Ferns and allies	164	13
Resident birds	94	40
Land mammals	19	95
Reptiles	13	100
Amphibians	4	100
Freshwater fish	25	76
Arthropods	~10 160	~20

found there are invaders, as are virtually all of the mammals (Table 11.1). New Zealand has also been particularly prone to invasion. Well over 50% of the flora of New Zealand is now composed of invaders. Unfortunately, the extinction of native species often accompanies such invasions.

11.4 OLD WORLD ECOSYSTEMS

One important contribution of the SCOPE program was an analysis of the numbers of invaders into nature preserves (Usher *et al.*, 1988). It was found that nature preserves on islands, even though they are managed for minimal disturbance, are composed of as many as 30% invading species among the resident plants and 18% vertebrate invaders (Table 11.2).

Invasions into preserves where the climate is more severe, such as arid zones, have proportionately fewer invaders. There is some evidence though that it is not climatic severity *per se* that serves as a barrier to invasions. Opportunity of introduction plays a critical role. Usher (1988) notes that the numbers of invaders that are found in preserves is related to the numbers of visitors that frequent them. High visitation brings with it a certain amount

Table 11.2 Mean percentages of introduced invasive species over total species present in nature preserves distributed world-wide in different habitat types. (From Usher, 1988)

Habitat	Vascular plants	Vertebrates
Islands	30.4	17.9
Mediterranean-type	12.3	4.2
Arid lands	9.2	2.3
Dry woodland and savanna	5.4	0.8

of disturbance as well as an increased probability of inadvertent introductions of alien organisms. Reserves in climatically challenging environments are evidently less heavily used by humans.

11.5 THE TRANSPORT OF INVADING SPECIES

Many invading species could not have been introduced without the help of humans, both intentionally as well as inadvertently. Intentional introductions include organisms that have been introduced to enhance agriculture and horticulture, to provide new forms of animal protein, and to develop recreation. In earlier times, in New World regions, acclimatization societies were formed with the goal of introducing and establishing organisms from the Old World. There are of course still efforts to introduce new plant varieties and game birds and fishes into regions where they have not previously existed. These activities have resulted in literally thousands of species introductions. At the same time there have been organized efforts to stem the flow of inadvertent introductions, principally pests of economically important plants. Plant quarantine activities have stemmed, but have by no means stopped, these inadvertent introductions. Then, too, as noted below, seemingly benign and beneficial plants can become 'weedy' and cause considerable economic damage in a new environment.

11.6 THE ORIGINS OF INVADING SPECIES

If one chooses any particular spot on earth and examines the origins of invading species, one finds that the invaders often come from radically different biogeographic realms. However, there can be significant patterns and directions of flow between certain regions. In the New World, for example, most invaders are of European origin. Is this unusually high success of European species due to some particular properties that they possess, perhaps resulting from their long association with the humans and subsequent exposure to disturbance (di Castri, 1988)? Or does this success simply reflect, in large part, opportunity for invasion due to the intensity and directionality of introductions (Simberloff, 1988)? In the New World there has been a constant flow of intentional introductions from the European and Middle Eastern centers of domestication of plants and animals. No doubt in the years ahead the patterns of biotic invasions will further change, due to changing cultural practices and transportation modes and networks, as has occurred in the past (di Castri, 1988).

11.7 THE IMPACT OF INVADERS

The economic and sociopolitical impacts of invading species have been unquestionably enormous. On the plus side one need only consider the major crop plants of the world that represent intentional and generally highly successful introductions. These species have provided the world's population with food, fiber, and fuel. Such crops, however, represent only a fraction of the species that have been intensively selected for desirable economic properties. Often this selection has taken place under monospecific conditions, and in the process such species have generally lost competitive ability and thus are no match, when not cultivated, with wild species. In contrast to the small number of major crop species there are over eight thousand weeds of agriculture, 250 of them being serious agricultural pests (Heywood, 1988). These weeds have had a very large economic impact that has been well documented.

What has not been so well documented is the impact that invaders, either intentionally or accidently introduced, have had on natural ecological systems. The SCOPE program focused on these natural systems and found examples of disruptions of virtually every ecosystem process including acceleration of erosion rates, alteration of biogeochemical cycling, alteration of geomorphological and hydrological processes, alteration of fire regimes, interference with the population dynamics of native species, and changes in community structure (Macdonald *et al.*, 1988). Vitousek (1986) noted that it was invading species, representing new functional roles within a given ecosystem, such as tapping new water sources or fixing nitrogen, that did the most damage. He and others (Vitousek *et al.*, 1987) have provided a recent example of a nitrogen-fixing shrub from the Canary Islands which has invaded the lava flows of Hawaii. This species has completely altered normal soil development and community successional patterns.

11.8 PREDICTING THE SUCCESS OF INVADING SPECIES

In spite of the wealth of information that we now have on the numbers, origins, and impacts of invading species we are still unable to make precise predictions about the success a given species will have in invading a particular system. We can, however, make some important general predictions. The reasons that we lack the kind of precision in predictability that we would like rests both in the complexity of the problem and in the nature of the data that we now have available. In regard to the latter most of the case histories that we have on invasions represent natural historical accounts, which generally do not provide details on quantitative population attributes such as population growth rates, rates of spread, etc. Further, there generally is not specific information available on failure rates of introductions with which to compare to successes, with some notable exceptions, e.g. data sets of Hawaiian birds (Moulton and Pimm, 1986). Finally, and probably most important, this area of study has not yet benefited from an experimental approach.

In addition to the inadequate data available on this problem are the formidable problems associated with making predictions. First, it is important to realize that there are many steps to the success of an invading species, each with numerous and different constraints. There is transport, establishment, reproduction, and, finally, spread. Thus any approach to predicting invasion success must take all the factors limiting steps into consideration. Second, ecological communities do not all behave in the same way; indeed, such systems are exceedingly complex involving processes that occur as quickly as milliseconds to those that occur on evolutionary time scales.

Several approaches have been taken toward developing the ability to predict, at some level, basic invasion processes. The first approach has been essentially correlative and comparative. This approach has sought to predict the potential success of an invading species based on morphological, physiological, ecological, and genetic characteristics which are associated with particularly successful invading species. Examples of lists of traits associated with successful species are those published for plants (Baker, 1974) and vertebrates (Ehrlich, 1988). Currently, this approach has been unable to make specific predictions. Lists of species traits are simply too general to be of value in specific instances. As noted by Ehrlich, other factors must be taken into account. These include the particular physiological and behavioral status of the founding population as well as the unique physical and biological status of the habitat being invaded (Table 11.3). Clearly, a species could be a good invader in one ecological system but a poor invader in another; conditions are rarely equal.

Even invasions by the same species into the same environment may not even result in the same outcome. Dynamical as well as stochastic processes can produce numerous alternative community states. Depending on the state of the system at the time of invasion, quantitative differences in vulnerability and resistance to invasion can occur. Even in the absence of different system states, timing can play a powerful role in invasion processes (Crawley, 1988). Attempt after attempt to establish an organism in a new environment may fail. Then a chance juxtaposition of climatic events, status of the populations of potentially interactive species, and the conditions and numbers of individuals of the invading species may result in a successful establishment. The large numbers of elements that go into a successful invasion make it striking indeed that any species actually becomes an established invader.

Considering the large numbers of crucial variables encompassed in these two realms of species traits and habitat characteristics, plus the potentially large numbers of species that are promising invaders for which we have

Table 11.3 Generalized features of populations and of communities important in the establishment of invertebrate invaders. (From Ehrlich, 1988)

A. Population features
1. Number of individuals
2. Sex or sex ratio
3. Physiological status—maturity, state in breeding cycle, health, acclimation status
4. Genetic composition—genetic variability in founding population, ecotype of origin
5. Behavioral status—experience of individuals, social relationships within group

B. Habitat and community features
1. Season
2. Weather
3. Size and structure of populations of resource organisms
4. Size and structure of populations of competitors
5. Size and structure of populations of predators
6. Size and structure of populations of parasites and pathogens

relatively little detailed biological information, the problem in making precise predictions becomes evident. Although it may be difficult to make *a priori* predictions about the likelihood of establishment and spread of any particular organism, without detailed knowledge of that organism's biology and the nature of the target habitat, such studies have produced some very useful guidelines.

Crawley (1987) has noted that species with large distributional ranges, a high intrinsic population growth rate, and a large founding population are most likely to be successful invaders. Communities with few species present, absence of congeners morphologically similar to potential invaders (Moulton and Pimm, 1986), and a high degree of habitat disturbance (Crawley, 1988) are most likely to be invaded. The greatest impacts of invaders on natural systems will be seen where predators and competitors of the invader are absent in the target community and in communities that have a relatively simple structure (e.g. those with few species). Invading species that are generalists, such as generalist herbivores, are likely to have a greater impact than those with specialized food requirements (Pimm, 1987).

Recent studies on the dynamics of community assembly provide further promise of delimiting those species which are better colonists and those ecological systems which are particularly vulnerable (e.g. Gilpin *et al.*, 1986; Drake 1983, 1988; Sugihara 1985; Post and Pimm, 1983). Using mathematical models, both Sugihara (1983, 1985) and Drake (1983) have shown that as communities become increasingly species rich, successful invaders must become increasingly specialized (diet, habitat use, etc.). Hence, the link between species and community characteristics appears even more critical if we are to develop predictive ability.

A conclusion, then, is that with detailed knowledge of the biology of the potential invader and of the characteristics of the target environment, relatively good predictions can be made of the potential success and impact of an invading species. Obtaining this knowledge for a specific system will doubtless have a price. The costs in manpower in research could be large, but then the costs of a mistake will inevitably be larger.

An important lesson of the utility of prerelease in-depth studies can be learned from the results of biocontrol efforts. Over 180 species of insect herbivores have been utilized in weed control efforts. About two-thirds of these became established and about 50% were effective in control. These introductions have resulted in some species inadvertently becoming established; however, none have themselves become a problem (Moran *et al.*, 1986).

11.9 AN ENVIRONMENTALIST'S VIEWPOINT

It is important to understand why environmentalists have expressed such strong concerns over the potential hazards of the introduction of genetically designed organisms. They are acutely aware of those cases where invasive species, inadvertently introduced, have disrupted natural systems or have caused enormous economic damage to crops. They can also cite many cases where species that have been purposefully introduced to serve some useful function for society have subsequently caused more damage than providing good. These latter cases represent an engineering approach to the environment in a sense that one part of the problem (e.g. economic benefit, enhancement of yield, erosion control, etc.) is considered in detail whereas other impacts or hazards are not considered.

Examples of such an approach in the United States must surely be the purposeful introduction of crabgrass and dandelions, the scourge of gardeners everywhere. One can say, quite correctly, that these particular introductions were made before we had a sophisticated understanding of the potential hazards that might ensue. However, even in relatively recent times, examples can be cited where purposeful introductions have been made with very good intent and expectations. In 1957, for example, the Nile perch was introduced into Lake Victoria. In this case the intent was to increase human food supply. Apparently even this objective was not fully realized, yet considerable ecological damage ensued including the apparent extinction of numerous endemic fish species (Barel *et al.*, 1985). It is no wonder, then, that even the words employed to describe the process of utilizing genetically modified organisms in natural environments—'the release of genetically engineered organisms'—conjures images to some biologists of yet one more well-intentioned but potentially damaging activity. Many engineering projects in natural environments have had, in past times, at least before the era of environmental impact considerations, rather devastating ecological effects. A combination of an engineering approach and releases of organisms thus brings double fears to environmentalists.

Despite such fears we have learned much since the days when species were intentionally introduced without the slightest consideration of the ultimate consequences. We must ask whether or not our new knowledge is adequate to prevent the catastrophies of history? We must be willing to evaluate the long-term consequences of our activities.

11.10 TODAY'S ENVIRONMENT

At present, there are drastically different configurations of the world's biota than there were even a few hundred years ago. We have seen an homogenization of the world's biota, as well as an enormous loss of species. These changes have accompanied, and have partly resulted from, large-scale modifications of pristine landscapes by human activities including agriculture, animal husbandry, water diversions, construction, and so forth.

As the human population grows there will be increasing impacts on the earth's biota, both directly, through the activities listed above, as well as indirectly, through the effects of increasing global interconnectivity. Of course it would be a mistake to assume that any cost associated with human activities that modify the Earth are unacceptable. It has been suggested that the inadvertent escapes among these intentional introductions represent, in a sense, a tax on the good that has been provided (Wells *et al.*, 1986). The damage that many of the escapes has wrought has been considerable; however, cost versus benefit is an ecological reality.

Do we now have the knowledge to more precisely utilize purposeful introductions without incurring environmental damage or to prevent inadvertent introductions that may also do damage? To a certain extent yes, as noted above, but unfortunately this knowledge is not often fully applied. Clearly, ecologists and molecular biologists must work together in the development, testing, and application of engineered species for use in the open environment.

11.11 SUMMARY

There are certainly lessons from the study of biological invasions that are of value in considering the deliberate introduction of engineered organisms. First, a detailed evaluation of traits of the organism and of the properties of the host environment are required before release. This analysis must be made in an ecological context that considers both the short- and long-term consequences of the introduction since it is clear that once an introduction is successful there will be no easy recall. If the history of introductions serves as a guide, we can predict that although there are great potential benefits to be derived from the considered utilization of biological materials in a given area, no matter what its origins, there will be mistakes due to unforeseen circumstances. As noted in the introduction, such mistakes will be minimal with the controlled introduction of engineered organisms since in most cases this will not involve moving organisms from one ecological setting into an entirely different one. It would appear that the potential benefits of engineeered organisms are enormous, not only in enhancing agricultural productive capacity but also in providing new tools with which to restore properties of natural ecosystems that have been inadvertently damaged through the activities of humans.

REFERENCES

Baker, H.G. (1974) Evolution of weeds. *Ann. Rev. Ecol. Syst.* **5**, 1-24.

Barel, C.D.N., Dorit, R., Greenwood P.H., Fryer, G., Hughes, N., Jackson P.B.N., Kawanabe, H., Lowe-McConnell, R.H., Nagoshi, M., Ribbink, A.J., Trewavas, E., Witle, F. and Yamaoka, K. (1985) Destruction of fisheries in Africa's lakes. *Nature* **315**, 19–20.

Crawley, M.J. (1987) What makes a community invasible? In: Gray, A.J., Crawley, M.J. and Edwards, P.J. (Eds.) *Colonization, Succession and Stability*, Blackwell Scientific Publications, Oxford, pp. 429–453.

Crawley, M.J. (1988) Chance and timing in biological invasions. In: Drake, J., diCastri, F., Groves, R., Kruger, F., Mooney, H., Rejmanek, M. and Williamson M. (Eds.) *Biological Invasions: A Global Perspective*, John Wiley, Chichester.

di Castri, F. (1988) History of biological invasions with special emphasis on the old world. In: Drake, J., di Castri, F., Groves, R., Kruger, F., Mooney, H.A., Rejmanek, M. and Williamson, M. (Eds.) *Biological Invasions: A Global Perspective*. John Wiley. Chichester.

Drake, J.A. (1983) Invasibility in Lotka–Volterra interaction webs. In DeAngelis, D., Post, W.M., and Sugihara, G. (Eds.) *Current Trends in Food Web Theory*, Oak Ridge National Lab. TM5983, pp. 83–90.

Drake, J.A. (1990) Communities as assembled structures: do rules govern pattern? *Trends in Ecology and Evolution* **5**, 159–164.

Drake, J.A. diCastri, F., Groves, R., Kruger, F., Mooney, H., Rejmanek, A., and Williamson, M. (Eds.) (1988) *Biological Invasions: A Global Perspective*, John Wiley, Chichester.

Drake, J.A. and Williamson, M.H. (1986) Invasions of Natural Communities. *Nature* **319**, 718–719.

Ehrlich, P.R. (1988) Attributes of invaders and invading processes. In: Drake, J., diCastri, F., Groves, R., Kruger, F., Mooney, H., Rejmanek, A., and Williamson, M. (Eds.) *Biological Invasions: A Global Perspective*, John Wiley, Chichester.

Gilpin, M.E., Carpenter, M.P., and Pomerantz, M. (1986) The assembly of a laboratory community: multispecies competition in Drosophila. In Diamond, J. and Case, T.J. (Eds.) *Community Ecology*, pp. 23–40, Harper and Row, New York.

Groves, R.H. (1986) In: Groves, R.H. and Burdon, J.J. (Eds.) *Ecology of Biological Invasions: An Australian Perspective*, Australian Academy of Science, Canberra, pp. 137–49.

Groves, R.H. and Burdon, J.J. (Eds.) (1986) *Ecology of Biological Invasions*, Australian Academy of Science, Canberra.

Heywood, V.H. (1988) Patterns, extents and modes of invasions by terrestrial plants. In: Drake, J., diCastri, F., Groves, R., Kruger, F., Mooney, H.A., Rejmanek, M. and Williamson, M. (Eds.) *Biological Invasions: A Global Perspective* John Wiley, Chichester.

Joenje, W., Bakker, K. and Vlijm, L. (Eds.) (1987) *Proceedings*, Series C, Koninklijke Nederlandse Akademie van Wetenschappen, Vol. 90 (1).

Kornberg, H. and Williamson, M.H. (Eds.) (1986) Quantitative Aspects of the Ecology of Biological Invasions. *Phil. Trans. R. Soc. Lond. Ser. B*, **314**.

Macdonald, I.A.W., Kruger, F.J. and Ferrar, A.A. (1986) *The Ecology and Management of Biological Invasions in Southern Africa*, Oxford University Press, Cape Town, 324 pp.

Macdonald, I.A.W., Loope, L.L., Usher, M.B. and Hamann, O. (1988) Wildlife conservation and the invasion of nature reserves by alien species: a global perspective. In: Drake, J., diCastri, F., Groves, R., Kruger, F., Mooney, H., Rejmanek, M. and Williamson, M. (Eds.) *Biological Invasions: A Global Perspective*, John Wiley, Chichester.

Mooney, H.A. and Drake, J.A. (Eds.) (1986) *Ecology of Biological Invasions of North America and Hawaii*, Ecological Studies 58, Springer Verlag, New York.

Moran, V.C., Neser, S. and Hoffmann, J.H. (1986) The potential of insect herbivores for the biological control of invasive plants in South Africa. In: Macdonald, I.A.W., Kruger, F.J. and Ferrar, A.A. (Eds.) *The Ecology and Management of Biological Invasions in Southern Africa*, Oxford University Press, Cape Town, pp. 261–80.

Moulton, M.P. and Pimm, S.L. (1986) Species introductions to Hawaii. In: Mooney, H.A. and Drake, J.A. (Eds.) *Biological Invasions of North America and Hawaii*, Ecological Studies 58, Springer Verlag, New York, pp. 231–48.

Pimm, S.L. (1987) Determining the effects of introduced species. *Trends in Ecology and Evolution* **2**, 106–8.

Post, W.M. and Pimm S.L. (1983) Community assembly and food web stability. *Math. Biosci.* **64**, 169–92.

Sailer, R.I. (1983) History of insect introductions. In: *Exotic Plant Pests and North American Agriculture*, Academic Press, New York, pp. 15–38.

Simberloff, D. (1988) Which insect introductions succeed and which fail? In: Drake, J., diCastri, F., Groves, R., Kruger, F., Mooney, H.A., Rejmanek, M. and Williamson, M. (Eds.) *Biological Invasions: A Global Perspective*, John Wiley, Chichester.

Stehli, F.G. and Webb, S.D. (1985) *The Great American Biotic Interchange*, Plenum Press, New York, 532 pp.

Sugihara, G. (1983) Holes in niche space: a derived assembly rule and its relation to intervality. In: DeAngelis, D., Post, W.M. and Sugihara, G. (Eds.) *Current Trends in Food Web Theory*, Oak Ridge National Lab. TM5983, pp. 25–33.

Sugihara, G. (1985) Graph theory, homology, and food webs. *Symp. in Appl. Math.* **30**, 83–101.

Usher, M.B. (1988) Biological invasions into nature reserves: a search for generalizations. *Biological Conservation* **44**, 119–135.

Usher, M.B., Kruger, F.J., Macdonald, I.A.W., Loope, L.L. and Brockie, R.E.

(1988) The ecology of biological invasions into nature reserves: an introduction. *Biological Conservation* **44**, 1–8.

Vitousek, P.M. (1986) Biological invasions and ecosystem properties: can species make a difference? In: Mooney, H.A. and Drake, J.A. (Eds.) *Ecology of Biological Invasions of North America and Hawaii*, pp. 163–176. Springer Verlag, New York.

Vitousek, P.M., Loope, L.L. and Stone, C.P. (1987) Introduced species in Hawaii: biological effects and opportunities for ecological research. *Trends in Ecology and Evolution* **9**, 224–7.

Wells, M.J., Poynton, R.J., Balsinhas, A.A., Musil, K.J., Loffe, H., van Hoegsen, E. and Abbott, S.K. (1986) The history of introductions of invasive alien plants to southern Africa. In: Macdonald, I.A.W., Kruger, F.J. and Ferrar, A.A. (Eds.) *The Ecology and Management of Biological Invasions in Southern Africa*, Oxford University Press, Cape Town, pp. 21–35.

Section 3

ASSESSMENT AND REGULATION

Introduction of Genetically Modified Organisms into the Environment. Edited by Harold A. Mooney and Giorgio Bernardi

12 The Ecology of Genetically Engineered Organisms: Assessing the Environmental Risks

MICHAEL J. CRAWLEY
Department of Pure and Applied Biology, Imperial College, Ascot, Berkshire, UK

12.1 ECOLOGICAL BACKGROUND

There seems to be a view that the ecology of genetically engineered organisms is somehow different—more inscrutable perhaps, but certainly more dangerous—and that the intentional release of genetically engineered organisms poses more of a threat to the balance of nature than other kinds of organisms bred by man. This view is mistaken. While there are risks associated with the introduction of any novel organisms into a habitat, the ecology of genetically engineered organisms is exactly the same as the ecology of any other living thing. The rules are precisely the same, no matter how the genotype is put together. Populations must have an intrinsic rate of increase greater than zero in order to persist (see below). They must be supplied with essential resources at a sufficient rate to allow this multiplication rate to be expressed. Transgenic individuals are exposed to predators, parasites, diseases, and competitors, and suffer the same kinds of losses during dispersal as any other organisms. They may require mutualists for resource gathering, reproduction, defence, or dispersal. They must deal with the vagaries of changing weather and heterogeneous substrate in the same way as any other organisms.

It is natural, therefore, to ask whether the risks of introducing transgenic and non-transgenic organisms are significantly different. The track record of non-transgenic introductions has been mixed. Most of the world's major crops and domestic animals are introduced, and have been transported all over the world to man's evident benefit, without adverse environmental repercussions. A number of introductions, however, have gone seriously

Introduction of Genetically Modified Organisms into the Environment. Edited by Harold A. Mooney and Giorgio Bernardi

wrong (e.g. goats on islands; salvinia fern in tropical waterways; cane toads in Australia; see Williamson, 1988).

We need to know whether or not the occasional ecological disasters that have followed from unplanned species' introductions represent a good model for the risks associated with the deliberate introduction of transgenic organisms. First, we should distinguish between the introduction of modified genotypes of existing, long-cultivated organisms, and the introduction of brand new, alien forms (whether or not these are genetically engineered). It is likely that the risks are substantially different for these two categories of introduction; for example the risks associated with transgenic maize in the US are probably low, whereas the risks associated with introducing a nitrogen-fixing tree to the Hawaiian Islands are substantial (the risks of introducing a transgenic, nitrogen-fixing tree may well be higher still).

Most of the evidence from existing studies suggests that the release of genetically engineered crop plants will be safe (Bradshaw, 1984; Davies and Snaydon, 1976; Hedrick, 1986; NAS, 1987). Conventional wisdom is based on four notions: (1) almost all genetic change reduces fitness; (2) wild-type genotypes are almost always competitively superior to introduced genotypes of the same species; (3) there is no 'free lunch' in evolution—all advantages accruing from the engineered genes must be paid for, and this will lead to reduced fitness in the wild where, in the absence of selection in favour of the trait, the engineered gene is more likely to be a liability than a benefit; (4) however, some engineered traits (e.g. increased tolerances to extreme environmental conditions, or increased resistance to insect pests or fungal pathogens) may be advantageous under natural field conditions, and these traits are the most likely to prosper away from the cosseting of arable agriculture.

12.2 PHENOTYPE AND GENOTYPE

Evidence on the performance of phenotypes expressed by different genotypes in natural ecosystems is limited. The best information is obtained from studies where genotypes are subject to replicated, reciprocal transplant experiments into a range of plant communities. It is unfortunate that in early studies of ecological genetics, the transplants were grown only in common gardens, where they were freed from competition with native vegetation and protected from herbivores by fencing (e.g. Clausen *et al.*, 1940). These studies were immensely influential, however, because they showed that there was substantial genetic variability for morphological traits within a species. To understand the population biology of introduced genotypes, it is necessary to know how the various genotypes would fare in comparison with resident types, when confronted with the full rigours of competition, predation, disease, and abiotic extremes to which they would be exposed in native vegetation.

In most cases where introduced plant genotypes have been grown in competition with native vegetation, the relative fitness of the alien genotype has been shown to be substantially lower than the wild type. Levin (1984) sowed 1100 seeds of alien genotypes of the annual *Phlox drummondii* into natural populations in 1973. In 1974, 89 of the alien genotypes flowered, and plants continued to appear from this sowing until 1979 as a result of seed dormancy. The alien population was extinct by 1981. Working with the buttercup *Ranunculus repens*, Lovett Doust (1981) found that short-distance reciprocal transplants between woodland and grassland populations led to relative fitness values of about 0.2 for the alien genotype. Transplanting *Plantago major* between lawns and roadsides, Warwick and Briggs (1980) obtained similar results. Some of the most detailed transplant studies were carried out by Davies and Snaydon (1976) using genotypes obtained from different treatments of the Park Grass Experiment, begun at Rothamsted in 1856. Cloned ramets of *Anthoxanthum odoratum* performed significantly less well when transplanted back into plots with an alien fertilizer regime than when transplanted back into the sward from which the material originated. Genetic differences could be detected in response to liming and fertilizer treatments which had been begun as little as 7 years previously. When species-rich grassland swards are sown from seed, a combination of grazing susceptibility and competition from grazing-tolerant plants usually leads to a rapid reduction in species (genotype) diversity (Charles, 1961; Snaydon, 1978).

Similar results have been obtained with mixtures of crop plant genotypes. For example, Harlan and Martini (1938) grew ll varieties of barley at 10 experimental stations at roughly the same latitude across the United States. Within as little as four years, most of the varieties were close to extinction at most of the sites. Different genotypes, however, achieved dominance at different locations. Jennings and Jesus (1968) carried out similar studies with rice, and found rapid reductions in genotype diversity. Their study is particularly interesting for the fact that the identity of the surviving genotype could not be predicted on the basis of performance of the genotype when grown alone. Clearly, selection against introduced genotypes acts both quickly and intensely under field conditions.

These findings present us with something of a dilemma. Experiments suggest that selection rapidly eliminates introduced genotypes from natural populations, and yet apparently homogeneous natural communities contain numerous genotypes, often spectacularly many (Gottlieb, 1981; Hutchings, 1986). How we account for the maintenance of this high level of genetic variability, in the face of the evidence showing rapid competitive exclusion of alien genotypes, is one of the major unanswered questions of ecological genetics (Bradshaw, 1984; Loveless and Hamrick, 1984; Hedrick, 1986).

There are two obvious explanations for this. The first is that most genetic change is neutral rather than deleterious. The second is that real environments are not uniform, and it is environmental heterogeneity that maintains the high levels of genetic variability observed in the field.

12.2.1 THE RELATIVE FITNESS OF ENGINEERED GENOTYPES: THE COMPLACENT VIEW

Conventional wisdom, based on laboratory studies and on the results of the transplant experiments reviewed above, suggests that most genetic changes are deleterious. This leads to an optimistic view for the prospects of safety in releasing genetically engineered organisms. Their genotype has been altered, so they will be less fit, and therefore they will die out under field conditions because they are 'less competitive'. A symptom of this view is a tendency to speak of a transgenic plant's 'competitive ability' as if this were some kind of genotype-specific trait. This is a mistake. Ecological performance is highly context specific, and the same genotype will give rise to phenotypes with different fitnesses in different environments.

12.2.2 THE RELATIVE FITNESS OF ENGINEERED GENOTYPES: THE CIRCUMSPECT VIEW

While it is true that no carefully controlled field study has ever shown the replacement of a wild-type genotype by an introduced alien genotype of the same species, it is equally plain that wild types are not the 'best' in any absolute sense. They may be the best (i.e. those of highest fitness) out of the set of genotypes currently available in a given habitat. Alternatively, existing genotypes may represent those that were fittest at the time when the present cohort of individuals became established. Such genotypes might not be fittest if recruitment were to occur now, because environmental conditions could have changed in the meantime (especially if the organisms are long lived). In any event, there will always be the potential for genotypes of higher fitness than the wild type (present-day, wild-type plants are not 'perfectly adapted to their environment'; see Harper, 1982). When such superior genotypes arise, they will increase in frequency as a result of evolution by natural selection. Superior genotypes could be produced either accidentally or intentionally by genetic engineering. It is for this reason that a degree of circumspection is required in assessing the risks associated with the introduction of transgenic crop plants; there is no cause for alarm, but prudence requires that we proceed cautiously.

12.2.3 VARIATION FOR FITNESS TRAITS

It is important to establish how much of the genetic variation observed in natural populations relates to fitness traits, and the extent to which the ecological performance of phenotypes is affected by differences in genotype. The evidence here is equivocal. In many cases, there appears to be little or no scope for selecting traits relating to development rate, survivorship, or fecundity (Bradshaw, 1984), and most of the genetic variation appears to be neutral (at least in existing environments, at the present time). In other cases one fitness attribute can be improved under experimental selection, but only at the expense of trade-offs in other parameters (Smith *et al.*, 1987).

The question therefore arises as to whether genetic engineering can ever give us something for nothing. Can the breeder introduce a desirable trait without altering other important attributes of yield or quality that might reduce fitness under field conditions? Is it inevitable that the engineered genotype will have a lower intrinsic rate of increase under field conditions than the genotype it is introduced to replace? I shall restrict my discussion to transgenic crop plants, although many of the ecological considerations are general and would apply equally well to transgenic microbes or domestic animals.

12.2.4 RISKS ASSOCIATED WITH TRANSGENIC CROP PLANTS

Commercial introduction of genetically engineered crop plants is likely to be contingent upon a thorough assessment of the potential risks involved. Risks are currently considered as falling into three categories: (1) persistence (the transgenic plants become serious arable weeds); (2) invasion (the transgenic crops become invasive of natural habitats); and (3) transfer of introduced genes via pollen to other plant species (such that these then become persistent or invasive weeds). Little detailed information is available to allow assessment of risks in any of these categories, and research is required to provide such data prior to commercial introduction of transgenic crop varieties. The risks are not currently perceived as being high; transfer of genes resulting from conventional crop breeding into non-crop plants has not created conspicuous problems; nor have traditional crop plants themselves become invasive of natural habitats (Williamson (1988) lists some introduced crops that have become weedy in tropical countries).

The aims of ecological risk assessment are: (1) to determine the potential for persistence and spread of transgenic crops in a variety of habitats; (2) to discover the range of plant species that can cross-pollinate with transgenic crops; (3) to investigate the ecological performance of hybrid plants produced by such pollination; and (4) to develop protocols that would allow crop breeders to carry out their own ecological risk assessments on new transgenic plants in the future.

12.3 THEORETICAL BACKGROUND

I shall expand a model of invasion (Crawley, 1986) in order to understand the circumstances under which a genetically engineered crop plant might become the 'perfect weed'. I shall refer to the intentionally released, genetically engineered plant as the invader and the resident plants as the wild type. The wild type may be genotypes of the same species, if the plant is already common in the environment, or they may be unrelated species of the same guild, in cases where the released species is new to the environment (for details see Crawley, 1986, 1987, 1988). In this thought-experiment, constraints become benefits and various limiting factors are circumvented. Armed with this picture of the perfect weed, we can look critically at the available evidence to assess the likelihood of this kind of problem coming about as a result of genetic engineering.

12.3.1 POPULATION BIOLOGY OF TRANSGENIC PLANTS

The model for invasiveness is:

The rate of increase of the transgenic plant in a given habitat
= plant development rate
+ its seed production (timing and duration)
+ survival of vegetative parts (discounted by their mortality rate)
− the effects of competition with other plants of the same kind
− the effects of competition with other plant species
− the effects of herbivores (molluscan, insect, and vertebrate)
− the effects of fungi and other plant diseases
− the effects of mutualists (if they are in short supply; pollinators, seed dispersal agents, mycorrhizal fungi, etc.)
+ immigration of transgenic seed from other sites
+ establishment of transgenic plants from dormant seeds in the soil (seed bank)

The parameters of the model determine whether the introduced genotype will increase or decrease. In terms of risk assessment, we are interested in the circumstances under which the rate of increase is positive. If the rate is positive, then we need to ask what level of abundance would be reached and what environmental consequences would follow from the plants persisting at this density.

12.3.2 THE INVADER HAS A GREATER INTRINSIC RATE OF INCREASE THAN WILD STOCK

A higher rate of increase would lead rapidly to the replacement of the wild type by the invader. The most obvious way for this to come about is for the invader to be more efficient than the wild type in converting resources into progeny: it may convert a higher proportion of its photosynthate into productive leaf tissue; it may have a higher rate of net photosynthesis per unit leaf area or a higher water-use efficiency; it may extract limiting soil nutrients more efficiently or produce a larger crop of seeds from the same productivity. In the absence of natural selection for the engineered trait, however, the invader could only possess an intrinsic rate of increase equal to that of the wild type, if there were no fitness cost associated with the engineered trait.

12.3.3 THE INVADER HAS A HIGHER THRESHOLD OF DENSITY DEPENDENCE

Populations with a positive intrinsic rate of increase can not increase indefinitely. At some point, their further increase is prevented by competition for limiting resources; the population density at which this occurs is known as the density threshold. Possession of a higher density threshold would allow the invader to continue to increase after the wild type had run out of resources or had come to equilibrium at the prevailing rate of resource supply. The most obvious means of increasing the size of the potential equilibrium population is by expanding the range of resources utilized. This may take the form of exploiting a wider range of resources or being able to live in a greater variety of habitats. Alternatively, the invader may process the same resources more efficiently than the wild type (Tilman, 1988). In both cases, however, the invader increases its population density by increasing the width of its niche relative to the wild type.

The invader may become more abundant than the wild type by circumventing density-dependent control, simply by becoming more tolerant of crowding (e.g. if the wild type is regulated at a density below that set by the rate of resource supply). This could result from its acquiring various kinds of tolerance to chemical interference (e.g. to allelopathic chemical produced by the wild type; see Rice, 1984).

12.3.4 LOWER MORTALITY FROM NATURAL ENEMIES

The rate of attack from predators, parasites, and diseases may be reduced if the engineered organism lacked a feeding stimulant or host recognition factor or contained a novel feeding deterrent or toxin. The invader may obtain further advantage if it were to benefit from some frequency-dependent protection when rare (e.g. failure of the resident enemies to form a search image for it).

12.3.5 HIGHER RATE OF MUTUALISM

The released genotype may enter into qualitatively new mutualistic relationships that allowed it to increase more rapidly or persist at higher equilibrium densities. Alternatively, it may be more efficient in deriving benefit from the prevailing level of mutualistic organisms or it may form mutualistic associations at a higher rate than the wild type. In short, the invader would prosper if it were to use a wider spectrum of mutualist genotypes or use existing mutualists more efficiently.

12.3.6 HIGHER INTRASPECIFIC COMPETITIVE ABILITY

The released genotype may cause the competitive exclusion of the wild type if either resource limitation or intraspecific interference were the key factor in the population dynamics of the wild types. Improved competitive ability may result from faster rates of growth amongst the engineered individuals, from higher efficiency in resource conversion, or from increased effectiveness in interference competition (e.g. the engineered form is more tolerant of crowding or produces more potent allelopathic chemicals).

12.3.7 HIGHER INTERSPECIFIC COMPETITIVE ABILITY

The invader may reduce the fitness of individuals of competing species (per interspecific encounter) more than does the wild type. This may not be relevant in terms of the invader's population dynamics, unless the invader is also competitively superior to the wild type; otherwise the extra resources made available will be exploited preferentially by the wild type. This produces the rather curious situation in which the wild type increases in abundance as a result of the greater interspecific competitive ability of the invader. The wild type would only increase in abundance, of course, if the invader reduced the fitness of individuals of the competing species in habitats that were accessible to the wild type; it is more plausible, perhaps, that the invader might prosper as a result of increased habitat range, in which case it would only interact with the wild type over a fraction of its new-found range.

12.3.8 BIGGER OR MORE VARIED RANGE OF REFUGES

The invader may be able to persist through inclement conditions (adverse weather, physical disturbance, disease outbreaks, predator population explosions, etc.) in higher densities than the wild type. This may come about either by using more kinds of refuges or by using the same refuges more efficiently.

12.3.9 HIGHER RATE OF DISPERSAL

The invader may be uncompetitive locally, but still superior to the wild type as a result of more efficient dispersal. This may allow it to escape from competitive exclusion. More protracted seed dormancy would allow the invader to 'escape in time', and, if environmental fluctuations were sufficiently severe, its gain in fitness from this strategy might be sufficient to outweigh the loss in reproduction brought about by delaying reproduction in the dormant cohort of phenotypes. In a constant environment, of course, increased dormancy would be maladaptive, and would serve only to reduce the invader's rate of increase relative to the wild type.

12.3.10 ECOLOGY WITHOUT TRADE-OFFS?

All these elements have a common theme, in that they allow the invader to get something for nothing: a broader niche with no loss of efficiency; a higher rate of increase with no loss of competitive ability; increased population density with no increase in susceptibility to natural enemies; and so on. A useful approach is therefore to ask explicitly 'is it possible that genetic engineering can operate without trade-offs'? Can certain elements of phenotypic performance be improved without concomitant reduction in others? The evidence is meager, but it all points in one direction.

Experience with traditional breeding methods has shown repeatedly that trade-offs are to be expected. For example, the spectacular increases in cereal yields achieved in the last two decades have come about not by increasing total shoot weight but by increasing the proportion of shoot weight made up by seeds (Riggs *et al.*, 1981). (Note, also, that a large fraction of the yield increase has been the result of increased fertilizer and pesticide use, coupled with earlier planting, and has rather little to do with breeding, except insofar as varieties have been selected that respond best to these cultural practices; see Snaydon, 1984)

12.4 ECOLOGICAL ASSESSMENT OF TRANSGENIC PLANTS

12.4.1 BACKGROUND

In Europe, research on genetically engineered strains of sugar beet, potato, and oil seed rape has reached the stage where questions of ecological and environmental impact have been raised. In the United States, similar questions have been raised, but the crops involved are different: tobacco, tomato, soy bean, and maize are likely to be the first transgenic crops available commercially. Concern about the potential risks of deliberate introduction of transgenic crops centres on four questions: (1) will it become invasive of natural habitats?, (2) will the modified plant become a weed of other crops?, (3) will it become a volunteer weed?, and (4) perhaps most seriously, will the modified genes be transferred to other plant species by pollen hybridization (or by some other process) such that these other plants become weeds or become invasive of natural habitats?

While existing evidence suggests that the risks attendant on the introduction of transgenic crops are small and manageable (NAS, 1987), nevertheless, the risks are real and attempts must be made to quantify them. There are reasons to continue to be cautious, but there is no cause for alarm. We should recognize, however, that public perception of the benefits and risks of biotechnology is as likely to influence future industrial developments as is formal risk assessment.

In scientific terms, the study of these risk assessments affords us a unique opportunity to compare the ecological performance of organisms whose genotypes vary in a precisely controlled way. For the first time, we can experimentally manipulate genotype and observe the behaviour of its phenotypes in a range of environments.

12.4.2 QUESTIONS

A number of important questions about the ecology of genetically engineered crops need to be addressed in assessing the risks associated with their widespread cultivation (see Table 12.1). The risk assessment studies need to consider the *fate* of the genetically engineered plants (and their pollen) and the *effects* of the introduction on the environment (i.e. on subsequent crops in the same fields, on adjacent crops, and in nearby natural habitats).

To do this, the questions are broken down into three categories:

1. Problems concerned with the *persistence* of the vegetative plant and its propagules in different kinds of environment
2. Problems relating to the *spread* of the plant by vegetative growth and by seed, in both arable fields and natural habitats
3. Problems involving the risks of *lateral spread of the engineered genes*, either by pollination of different plant species or by other means

The risk assessment aims to establish that the intrinsic rate of increase of the transgenic plant is negative or acceptably small.

12.4.3 SAMPLING METHODS

An important requirement of these ecological risk assessment studies is that they are seen to be of general relevance. Each individual study must contain a sufficiently large number of individual plants that failure of any plants to survive in a given experiment is a sound estimate of the risk of persistence.

Moreover, each treatment must be well replicated, both within sites and over a wide variety of sites. This range and replication must reassure potential critics that transgenics have been assessed in (1) a full range of arable conditions (this is relatively easy); (2) a reasonable range of natural habitats (this is more difficult, since exhaustive study is clearly impossible and sensible compromises will have to be made—a minimum set may include grassland, woodland, and an open, disturbed habitat, in order to obtain a range of plant cover, shading, substrate stability, and successional age); (3) a representative range of climates and soil types.

The different demographic parameters of the model would be assessed by rather different methods. In order to estimate fecundity and mortality schedules, for example, transgenic plants could be sown in factorial density combinations in, say, 2×2 m permanent quadrats, along with (1) non-transgenic crop individuals and (2) native vegetation. In studies of persistence, it would be necessary to bury seeds and vegetative parts (including root fragments, tubers, etc.) at a range of depths, in a variety of soils (differing in their wetness, aeration, frost risk, etc.). Dormancy characteristics would take several years to estimate and should first be established for non-transgenic material. With transgenic material, the risk of long-term dormancy bringing novel genetic material into unusual ecological circumstances at some time in the future can not be discounted. This risk, however, is extremely difficult to assess experimentally. Perhaps the most realistic option is to engineer unique 'labels' into transgenic material released into the field, so that if problems do arise later, at least the source of the material can be traced by sequencing to read the label.

12.4.4 POLLEN FLOW

For each crop, it would be necessary to carry out a detailed study of flowering phenology (the timing and amount of pollen production), the animal species that visit the flowers, the other plant species in the vicinity of crops (i.e. up to about 1000 m) that these pollen vectors visit (including their phenology of stigma receptivity), and the amounts and distances of pollen transported. There are four steps to this analysis: (1) estimate the probability that pollen will be moved to a given distance (and determine the factors influencing this distribution); (2) estimate the probability of pollination; (3) determine the probability that pollination will lead to the production of viable hybrid progeny; and (4) assess the ecological performance of any such hybrids. Studies using transgenic plants carrying, say, kanamycin resistance will allow screening for paternity in plants at a range of distances and compass directions (seeds can be germinated on kanamycin-bearing substrates). Because of the ease of mass screening germinating seeds from large numbers of maternal plants, this will allow, for the first time, a detailed, quantitative description of transgenic pollen movement.

Table 12.1 Detailed questions to be addressed in assessing the ecological risks associated with the commercial introduction of transgenic crop plants

Field observations	Field experiments	Contained experiments
1. Persistence		
What is the survival of the vegetative parts of the plant under a range of climatic conditions, on soils of different kinds and with different categories of drainage?	What is the fate of seeds sown into a range of plant communities, including other arable crops, forage crops, permanent grasslands, and natural habitats?	How is pollen viability affected in transgenic plants?
How is perennation affected by the introduced genes?	What is the fate of transplanted seedlings in different habitats?	How is seed dormancy affected?
What factors influence plant mortality outside arable fields and how are these influenced by the novel genes?	What is the fate of transplanted mature plants (or rootstocks) in different vegetation types?	How do transgenic plants perform in competition experiments with crop plants and with selected native plants?
What is the nature of seed dormancy under different environmental conditions, and how does the introduced genetic change influence triggering, duration, and hardiness during dormancy?	How long does experimentally planted seed remain dormant but viable in a range of soil types?	
2. Spread of the vegetative plant		
What is the seed production of the plant when grown in a crop and in natural vegetation?	What is the vegetative growth rate on different substrates and with different competing species?	Is seed size or morphology different in transgenic plants, and how might this affect seed dispersal?
Is seed production limited by the rate of pollination?	Is the thinning rule (i.e. density-dependent plant mortality) similar for transgenic and non-transgenic plants?	Do transgenic plants present greater risks of spread by vegetative fragments?
What is the germination rate of seeds in soil?	What kind of compensatory growth is exhibited (e.g. gap-filling)?	

What is the mortality of seeds and seedlings in arable soils and beneath native vegetation?

What is the phenology of seedling emergence and growth?

What are the natural enemies of the seedlings?

What is the role of vertebrate and invertebrate herbivores in crop and non-crop habitats?

What is the mechanism of seed dispersal?

How far are seeds dispersed and how does this vary with environmental conditions?

Do the seeds produced by plants grown outside arable fields give rise to a second generation of plants?

If the plant were to prove invasive, at what rate would it spread and which habitats would it occupy?

Which plant species (if any) are displaced when (and if) the plant is established in natural habitats?

Which plant species are responsible for the competitive suppression of the plant in different natural habitats?

Table 12.1 Continued

Field observations	Field experiments	Contained experiments
3. Horizontal gene transfer through pollen		
How much pollen is produced?	What is the fate of labelled pollen?	Which plant species allow pollen germination on their stigmas?
What is the phenology of pollen production and what is the phenology of stigma receptivity of other plant species growing in the neighbourhood of crops (i.e within 500–1000 m)?	How much pollen reaches the stigmas of other wild plants under different conditions?	How is pollen dispersal affected in transgenic plants?
	Which insects carry the pollen?	Which plant species form viable, hybrid seed and at what rate is this seed produced?
Over what distance is pollen dispersed under different meteorological conditions?	How far away from the crop can an individual, potted crop plant be pollinated and how does the rate of pollination fall off with distance under a range of habitat conditions?	What is the germination rate of hybrid seed?
Which insects visit the flowers, how much pollen do they carry, and which other plant species do they visit?		
Where is the pollen deposited, on which species, and in what numbers?	What plants make the most efficient 'pollen barriers' for the construction of guard rows; is it non-transgenic members of the same species or plants that form physical barriers to pollen flow or to insect flight?	What phenotypes are exhibited by hybrid individuals?
What is the geographic distribution of closely related wild plants in the vicinity of centres of crop cultivation and what is their small-scale (100's m) distribution as weeds within arable fields and on land adjoining field boundaries?		What is the performance of hybrid plants in competition experiments with crop plants and with selected native plants?
		What is the nature of perennation and vegetative dormancy in hybrid and transgenic plants?
What natural habitats are found within 1000 m of arable fields, in those areas where the crops are grown, and what flora is supported by these habitats?		

The next step involves a determination of the threat posed by transgenic hybrid plants, as weeds of crops and as invaders of natural habitats (see below). Contained greenhouse studies should attempt to create wide genetic crosses, by hand pollination with a range of related wild plant species. All related wild plants that grow within about 1 km of fields where the transgenic crop is likely to be grown should be screened. This would include all current weed species that may possibly become 'more weedy' as a result of obtaining engineered genes. Thus we shall know the probability of pollination (from field studies) and the probability of viable offspring being produced per pollination (from hand pollination). Hybrid plants produced by this means would then be analysed for their persistence and invasiveness in field trials.

12.4.5 PERSISTENCE

This question of persistence is addressed by sowing seed and transplanting seedlings, vegetative parts, and mature individuals of the unaltered crop plants into a range of natural communities differing in soil, drainage, and climate, in order to assess their ecological performance outside the setting of arable cultivation. Part of this work will involve the design of experimental protocols to define and to quantify 'competitive ability' as a guide to future risk assessments, where particular attention will be paid to questions of experimental design and necessary levels of replication.

Dormancy tests on the seeds and perennating vegetative parts of transgenic and hybrid plants should be undertaken in contained conditions in the first instance, although these will also need to be field tested in due course. It would be valuable, for example, to demonstrate prior to field testing that the engineered plants did not differ in their dormancy characteristics.

12.4.6 SPREAD

Field experiments on invasiveness will include: (1) reciprocal transplants into a range of locations and plant communities; (2) introduction at different population densities to establish whether there are threshold population sizes or 'founder population' effects; (3) estimation of seed production, viability, and seed-dispersal distances; (4) assessment of the impact of pathogens and vertebrate and invertebrate herbivores on plant growth and survival. The aim is to produce a list of the natural and crop habitats in which the transgenic crop would be capable of increasing in abundance were it to escape the confines of arable cultivation.

12.5 CONCLUSION

Release of genetically engineered organisms does not pose substantial *new* threats to the environment. Release of any non-native organism is potentially risky. The evidence clearly suggests, however, that releasing engineered crop plants is likely to be safe, especially if current practice for the release of potentially dangerous substances is reinforced (Johnson, 1982; Urban and Cook, 1986). Nevertheless, this optimistic prognosis has to be reconciled with widespread public concern about the environment. A conclusive, experimental demonstration of the absolute safety of transgenic organisms is impossible. However, if detailed, experimental assessments of the population biology of transgenic plants are carried out, before permission for commercial introduction is granted, then potentially problematical genotypes should be eliminated prior to release. At this stage, however, it is not clear how a simplified, inexpensive experimental protocol for ecological risk assessment will look. We shall need the experience gained during the kind of field trials described in this paper before any recommendations on protocol are possible.

Finally, Murphy's (safety of genetic releases) law should be borne in mind. This states that the biggest problems with genetically engineered organisms will come from those that look to be the 'safest' (i.e. from those cases where we perceive no risk at all). This is because, in cases where the organisms were perceived to be dangerous from the outset, such great care would be taken that problems would be spotted early and eradicated. The unforeseen rare event, which creates a novel kind of problem from a benign-looking crop, will be difficult to control because, once released, it is not practical to eliminate plants from all the habitats and from all the seed banks into which they have found their way.

12.6 SUMMARY

Some questions related to the release of genetically engineered organisms can be answered only with practical experience. Realistic, small-scale field tests are likely to be the only way potential risks from commercial scale uses of genetically engineered organisms can be evaluated. It should be emphasized, however, that these short-term studies are only appropriate to risk assessments on annual crop plants. Nevertheless, at the end of a three year study of the population biology of transgenic and non-transgenic annual crop plants, one should be in a position to:

1. Provide data on persistence and invasion in natural and arable habitats.
2. Show how (and if) genetic engineering alters these parameters.
3. Describe pollen spread by insect vectors and by other means.
4. Show how (and if) genetic engineering alters the production, spread, or compatibility of pollen.
5. Catalogue the wild plants that share insect pollinators with the crop.

6. Provide quantitative data on successful cross-pollination between the crop and its wild relatives.
7. Provide data on the persistence and invasiveness of any transgenic hybrid plants produced by crossing experiments.

Such data would provide sufficient information necessary to assess what ecological risks (if any) were associated with the commercial introduction of a transgenic crop. An additional risk assessment (not considered in this paper) would need to consider toxicological aspects of genetic engineering (i.e. the chemical composition of the products of the transgenic crops).

REFERENCES

Bradshaw, A.D. (1984) Ecological significance of genetic variation between populations. In: Dirzo, R. and Sarukhan, J. (Eds.) *Perspectives on Plant Population Ecology*, Sinauer, Sunderland, Mass., pp. 213–28.

Charles, A.H. (1961) Differential survival of cultivars of *Lolium, Dactylis* and *Phleum*. *J. Br. Grassland Soc.* **16**, 69–75.

Clausen, J., Keck, D.D. and Hiesey, S.M. (1940) Experimental studies on the nature of species. I. The effect of varied environments on Western North American plants. *Carnegie Institute of Washington, Publication* **520**, 1–452.

Crawley, M.J. (1986) The population biology of invaders. *Phil. Trans. R. Soc. Lond. Ser. B* **314**, 711–31.

Crawley, M.J. (1987) What makes a community invasible? In: Gray, A.J., Crawley, M.J. and Edwards, P.J. (Eds.) *Colonization, Succession and Stability*, Blackwell Scientific Publications, Oxford, pp. 429–53.

Crawley, M.J. (1988) Chance and timing in biological invasions. In: Drake, J.A. *et al.* (Eds.) *Biological Invasions*, John Wiley, New York, pp. 407–23.

Davies, M.S. and Snaydon, R.W. (1976) Rapid population differentiation in a mosaic environment. III. Measures of selection pressures. *Heredity* **36**, 59–66.

Edmunds, G.F. and Alstad, D.N. (1978) Coevolution in insect herbivores and conifers. *Science* **199**, 941–5.

Gottlieb, L.D. (1981) Electrophoretic evidence and plant populations. *Prog. Phytochem.* **7**, 1–45.

Harlan, H.V. and Martini, M.L. (1938) The effect of natural selection in a mixture of barley varieties. *J. Agric. Res.* **57**, 189–99.

Harper, J.L. (1982) After description. In: Newman, E.I. (Ed.) *The Plant Community as a Working Mechanism*, Blackwell Scientific Publications, Oxford, pp. 11–25.

Hedrick, P.W. (1986) Genetic polymorphism in heterogeneous environments: a decade later. *Annual Review of Ecology and Systematics* **17**, 535–66.

Hutchings, M.J. (1986) The structure of plant populations. In: Crawley, M.J. (Ed) *Plant Ecology*, Blackwell Scientific Publications, Oxford, pp. 97–136.

Jennings, P.R. and Jesus, J. de (1968) Studies on competition in rice. I. Competition in mixtures of varieties. *Evolution* **22**, 119–24.

Johnson, E.L. (1982) Risk assessment in an administrative agency. *The American Statistician* **36**, 232–9.

Levin, D.A. (1984) Immigration in plants: an exercise in the subjunctive. In: Dirzo, R. and Sarukhan, J. (Eds.) *Perspectives on Plant Population Ecology*, Sinauer, Sunderland, Mass., pp. 242–60.

Loveless, M.D. and Hamrick, J.L. (1984) Ecological determinants of genetic structure in plant populations. *Annual Review of Ecology and Systematics* **15**, 65–95.

Lovett Doust, L. (1981) Population dynamics and local specialization in a clonal plant (Ranunculus repens). II. The dynamics of leaves and a reciprocal transplant–replant experiment. *J. Ecology* **69**, 757–68.

NAS (1987) *Introduction of Recombinant DNA-Engineered Organisms into the Environment: Key Issues*, National Academy Press, Washington, DC.

Rice, E.L. (1984) *Allelopathy*, 2nd ed., Academic Press, London.

Riggs, T.J., Hanson, P.R., Start, N.D., Miles, D.M., Morgan, C.L. and Ford, M.A. (1981) Comparison of spring barley varieties grown in England and Wales between 1880 and 1980. *J. Agric. Sci.* **97**, 599–610.

Smith, R.H., Sibly, R.M. and Moller, H. (1987) Control of size and fecundity in *Pieris rapae*: towards a theory of butterfly life cycles. *J. Animal Ecology* **56**, 341–50.

Snaydon, R.W. (1978) Genetic change in pasture populations. In: Wilson, J.R. (Ed.) *Plant Relations in Pastures*, CSIRO, Australia, 253–69.

Snaydon, R.W. (1984) Plant demography in an agricultural context. In: Dirzo, R. and Sarukhan, J. (Eds.) *Perspectives on Plant Ecology*, Sinauer, Sunderland, Mass. pp. 389–407.

Tilman, D. (1988) *Plant Strategies and the Dynamics and Structure of Plant Communities*, Princeton University Press, New Jersey.

Urban, D.J. and Cook, N.J. (1986) Hazard evaluation division. Standard evaluation procedure. Ecological risk assessment. US Environmental Protection Agency, EPA 540/9-85-001. Washington, DC.

Wainhouse, D. and Howell, R.S. (1983) Intraspecific variation in beech scale populations and in susceptibility of their host *Fagus sylvatica. Ecological Entomology* **8**, 351–9.

Warwick, S.I. and Briggs, D. (1980) The genecology of lawn weeds. V. The adaptive significance of different growth habit in lawn and roadside populations of *Plantago major* L.*New Phytologist* **85**, 289–300.

Williamson, M. (1988) In: Drake, J.A. *et al.* (Eds.) *Biological Invasions*. pp. John Wiley, New York, pp. 329–50.

13 Ecological Issues Related to the Release of Genetically Modified Organisms into the Environment[1]

SIMON A. LEVIN
Charles A. Alexander Professor of Biological Sciences, Section of Ecology and Systematics, and Director, Center for Environmental Research, Cornell University, Ithaca, New York, USA

This paper had its genesis both in the Bellagio Meeting that led to the present volume and in the challenge of a joint special issue of *Trends in Ecology and Evolution* and *Trends in Biotechnology*, both published by Elsevier. The basic points to be made on both occasions were the same, but it seemed important to have the views expressed represented in both contexts. Hence, this chapter is, with only minor modification, a reprint of the paper 'Safety considerations for the environmental release of genetically engineered organisms', which appeared in *Trends in Biotechnology/Trends in Ecology and Evolution*. A similar version also appeared in *Chemical Technology and Biotechnology* published by Elsevier.

The number of recent meetings on the topic of the release of genetically engineered organisms is impressive. Unfortunately, the large number of meetings and opportunities to reprint this paper does not represent a rapidly expanding body of knowledge on potential risks and benefits. Most of what needs to be said of a general nature has been said long ago. What is needed now are approaches that abandon the generalities that have been prevalent and that deal with the specifics of individual cases and categories. A number of recent efforts have shown progress in that direction, and this volume is intended to be the first of a series that explore the potential for genetic engineering to assist in developing improved approaches to environmental management and ecological and evolutionary research.

[1] This article is reprinted from 'Planned release of genetically engineered organisms', in: Hodgson, J. and Sugden, A.M. (Eds.) *Trends in Biotechnology/Trends in Ecology and Evolution Special Publication*, pp. S47–S49, April 1988, with permission from Elsevier Publications, Cambridge.
Introduction of Genetically Modified Organisms into the Environment Edited by Harold A. Mooney and Giorgio Bernardi Published 1990 by John Wiley & Sons Ltd.

Introduction of Genetically Modified Organisms into the Environment. Edited by Harold A. Mooney and Giorgio Bernardi

The considerable debate that has been associated with proposals to introduce genetically engineered organisms into the environment obscures the fact that among most scientists there is strong basis for consensus concerning risks. Where disagreements do exist, they are concerned less with what the historical record says than with which part of it is most relevant to introductions of organisms developed with the new genetic techniques (Regal, 1988; Colwell, 1988). It is generally acknowledged that the overwhelming majority of introductions are likely to be benign, but that some will be problematical; nonetheless, the question can still be posed as to whether extrapolations should be based upon the average case, which will be low in risk, or upon the extreme high-risk cases, which will be low in probability.

Fortunately, such a simplistic question need not be answered, since it presupposes an artificial dichotomy. Most informed discussion recognizes that biotechnology holds tremendous potential for improving the human condition, including environmental applications that offer significant opportunities to reduce the pollutant load associated with the use of chemicals. There is general agreement that most introductions will pose minimal environmental risk and that thus there is no basis for generic arguments against the introduction of genetically modified organisms. On the other hand, there is no question that certain types of possible introductions pose a non-negligible degree of risk; thus, generic safety arguments must equally be rejected. Certainly, new genetic technologies will allow the creation of genotypes that never before have appeared and the introduction of genotypes into environments that are new for them. Thus, each potential introduction must be evaluated on its own merits. The inescapable conclusion is that some consideration is needed of the risks associated with the deliberate introduction of organisms, and furthermore that evaluation of risks must be specific to the particular application. This conclusion is universal in applicability. It is independent of whether or not one feels that existing legislation and procedures are adequate, an issue that requires different considerations for different nations.

13.1 WHY IS THERE DEBATE?

Given such strong elements of agreement regarding the basic issues, why has consensus been so long in developing? One reason, already suggested above, has been the inclination on all sides to generalize from special cases and the fact that the special cases of choice are not the same for everyone. Some refer to the long history of safe introductions associated with breeding in agriculture; others point to the problems that have been associated with the deliberate and accidental introductions of exotics. The relevance of these particular referents will be discussed below, but the differences of opinion and perspective point again to the necessity to get down to the specifics of individual cases and classes of introductions and to avoid broad generalizations.

A second and related reason for the debate reflects a difference in points of view. Some argue that the risks associated with the products of biotechnology are no different in kind than those associated with conventional procedures and that therefore no special attention is merited. Many who argue to the contrary accept the basic premise that there are no qualitative differences in risk, but emphasize the fact that there are risks in any introduction. They argue either for stricter controls on all introductions or that lax procedures regarding other types of introductions do not justify extending this laxity to an even broader class of introductions (Thompson, 1987).

A third reason that consensus has not been immediate is a specious philosophical dichotomy concerning who bears the burden of proof. Environmentalists point to the problems that often attend new technologies and argue that when new technologies are proposed the burden of proof regarding safety should rest upon the proposer (Levin and Harwell, 1986; Levin, 1986). On the other hand, proponents of biotechnology argue that the discussion of problems with deliberate introductions is hypothetical, dealing with the possible rather than the probable. They argue that there have been no examples of environmental problems associated with any introductions of recombinant-DNA organisms; indeed, only a handful of introductions have been carried out. Therefore, they conclude, the democratic process requires an assumption of innocence rather than guilt, and the burden of proof should rest upon those who construct unlikely scenarios for disaster.

Both of these lines of argument mislead. The fact of the matter is that we are not dealing with a new technology when we are dealing with the introduction of modified organisms into the environment. Humans have modified organisms genetically for centuries by a variety of techniques, and the repertoire of available tools has evolved and grown continually. Recombinant-DNA methods are the latest in the continuum of techniques, but do not represent anything inherently different from other advances that have occurred, except an ever-increased ability to carry out modifications efficiently and precisely. To the extent that increased efficiency leads ultimately to a change in the scale or frequency of introductions, that is worth consideration in terms of cumulative risks; indeed, it is this possible quantum increase in patterns of use that has been the major concern of many who have advocated caution. However, for any individual introduction, it is the characteristics of the engineered organism and its recipient environment that should be the basis for risk assessment, not the method by which the organism was modified (National Academy of Sciences, 1987). Thus, the appropriate line of argument, and one that leads much more rapidly to consensus, is as follows: We have sufficient understanding of the roles of genetics and environment in controlling the expressed phenotypic properties of an organism to allow us to state that it is not the method of modification that matters; what is relevant is the product and how it will be used. Therefore, as the recent US National Academy of Sciences report concludes, 'The risks associated with the introduction of recombinant-DNA-engineered organisms are the same in kind as those associated with the introduction into the environment of unmodified organisms and organisms modified by other genetic techniques' (National Academy of Sciences, 1987). This does not argue that those risks should be ignored, but simply that the considerations are no different in kind than those that apply for other kinds of introductions. Thus, we are not operating in an information vacuum, and have a large body of experience to draw on in assessing risks.

Although there is a growing recognition that it is more appropriate to focus on the product and patterns of use rather than on the method of production and that this should be the most important principle underlying regulatory procedures, current practice in the United States and other countries has not completely adjusted. Recently, for example, the National Institutes of Health concluded that a researcher at Montana State University, Gary Strobel, had not violated their guidelines when he introduced a genetically modified bacterium into the environment as part of an effort to combat Dutch elm disease. The explanation was that the introduced organism, a non-pathogenic *Pseudomonas syringae*, was the product of mating a second organism, which was recombinant-DNA modified, with a *Pseudomonas* bacterium that was not. The end product, it was adjudged, technically was not recombinant. This legal legerdemain is the inescapable consequence of having decisions made by a committee, the Recombinant Advisory Committee (RAC), whose very title is based on method rather than product. The irony of the whole situation is that Strobel's introduction of a known pathogen, the Dutch elm fungus, as part of the experiment, attracted virtually no attention because the fungus was not genetically engineered. Although Strobel took precautions to prevent the spread of the fungus, its introduction undoubtedly posed a greater hazard than did the introduction of the bacterium that was designed to control it, and a rational regulatory system should recognize that.

13.2 WHAT ARE THE CONCERNS AND THEIR RELEVANCE TO RISK ASSESSMENT?

13.2.1 INTRODUCED ALIEN SPECIES

One of the most compelling concerns for ecologists has emerged from their awareness of the problems that have sometimes attended the deliberate or accidental introduction of species taken from other environments. Charles

Elton (1958) in his classic text on biological invasions, was the first to bring focus to this problem. Increasingly, in a growing number of environments, invasions are being recognized as serious threats to the preservation of what we choose (by our choice of time scale) to regard as native fauna and flora (Mooney and Drake, 1986). Although the great majority of accidental introductions undoubtedly fail to become established (the exact statistics are impossible to determine, since we generally do not know about failed introductions), a substantial number do become established, and some of these become serious pests. Examples are legion: the European rabbit in Australia, the kudzu vine in the United States, the muskrat in Europe, the Africanized bee in South America, and forest and agricultural pests, as well as human pathogens and disease vectors, world-wide. Yet the relevance of these examples to most cases of introduction of genetically modified organisms has been severely challenged, because the transfer of a species from one environment to another, lacking its co-evolved biotic controls, involves fundamentally different considerations than does the reintroduction of a slightly modified species into the environment from which it was taken.

Most introductions currently under consideration fall into the category of reintroductions, but certainly there will be introductions proposed that involve alien species or that involve engineering to expand the range of some desirable species, such as popular sport fish. To address this, Levin and Harwell (1986) introduced four categories of introductions, and argued that separate considerations must apply to each. Some, and probably most, introductions will be reintroductions of native species; for such introductions, the alien species model is not a very good one, and conventional experience in agriculture provides a much more appropriate basis for extrapolation. The other two major categories involve nonnative species that are not found in the target environment because either (1) they cannot survive there without continual supplement or (2) they can survive there, but never before have been introduced. It is the latter category that encompasses the destructive examples mentioned earlier and presents the greatest cause for concern. The final category mentioned by Levin and Harwell is a catchall for the remaining cases, namely those that would be sufficiently novel to have no close analogues in any environment. For these cases, no good basis for extrapolation exists.

13.2.2 PATHOGENIC MICROORGANISMS

It is well recognized that small genetic changes in pathogenic species can convert avirulent types into virulent ones, or can permit expansions of the host range. Such small genetic changes underlie the outbreaks of new strains of human diseases such as influenza or of fungal pathogens such as the black stem rust of wheat. Thus, engineering of pathogenic organisms must be carried out with the utmost precaution, and with proper attention to safety. Similarly, if one were engineering nonpathogenic organisms with the intention of altering properties related to pathogenicity, similar care would be mandated. However, if nonpathogenic organisms are being altered with respect to properties unrelated to pathogenicity or invasibility, the likelihood of accidentally producing a pathogenic organism is negligible; pathogenicity involves not one but a complicated suite of characters (National Academy of Sciences, 1987). Much of the confusion on this point centers around the definition of a pathogen. To a pathologist, an avirulent strain of a pathogen is still a pathogen and should be handled with appropriate care. However, a single bacterial species, such as *Escherichia coli* or *Pseudomonas syringae*, may contain both pathogenic and nonpathogenic strains that, despite their common taxonomic identification, are genetically very different, and not interconvertible. The complexities of this issue and the subtleties and inadequacies of bacterial systematics have made it difficult to clarify the regulatory issues, and current practice is not yet totally satisfactory.

13.2.3 WEEDINESS

Introduced weedy species, such as *Melaleuca* or *Hydrilla* in the United States, are a familiar problem. However, weediness is highly unlikely to arise by accident when plants are being engineered for unrelated characteristics. A more substantial concern involves the possible exchange of genetic material between the domesticated targets of genetic engineering and their wild and weedy relatives. To evaluate this risk, one requires a detailed understanding of the ecological relationships in the particular environment. In the United States, for example, the potential for such exchange would be negligible for most row crops, although sorghum provides a striking counterexample. However, for forage grasses in the United States, the potential for introgression between wild and domesticated species varies widely (Mack, 1988). In South America the situation would be far more problematical for row crops: there is considerable exchange and introgression between domesticated crop species and wild and weedy relatives.

Properties such as herbicide resistance, if introduced into crop species, could well find their way into weedy populations if they impart a fitness advantage to the bearer. Such transfer could exacerbate chemical pollution problems if it led to the need to apply new herbicides to the weedy populations. It should be remarked that this is not a problem that has anything to do with the method of genetic engineering *per se*; once again, it is the properties of the introduced organism that are important, not how those properties were achieved. For any introduction of plants, whether of genetically modified or other organisms, what is needed is better information on the potential for exchange with wild and weedy relatives.

13.2.4 HORIZONTAL TRANSFER

Sexual exchange of material between domesticated plants and weedy relatives is but one of a variety of mechanisms of exchange known to occur among species. Bacterial populations, which usually do not exchange chromosomal DNA (Selander *et al.*, 1987), can exchange mobile genetic elements freely among species, and this has been important in the evolution of pathogenic traits (National Academy of Sciences, 1987). The most familiar example of such exchange involves the wide proliferation of antibiotic resistance among distinct bacterial species (Levy, 1982). Such transfer and spread is facilitated when there is selective pressure favoring it, for example when the traits borne on plasmids impart a selective advantage in polluted environments; thus the possibility for such spread must be given serious consideration when the environmental conditions are likely to favor the spread of the engineered trait. Although the frequency of horizontal exchange in nature is not established, such exchange is unlikely to be an important consideration in the absence of selective advantage favoring spread.

13.3 CODA

Ecological communities are dynamic assemblages, neither closed to invasion nor robust in the face of all perturbations. The relevant concerns of ecologists regarding the introduction of genetically engineered organisms into ecological communities are the same as would apply to organisms modified by any technique or to unmodified organisms. Increasing our ability to anticipate which introductions will be successful and which will lead to undesirable secondary effects is needed both for risk assessment and for developing better products.

The use of genetically engineered organisms holds considerable promise for environmental management and other purposes, provided appropriate safety standards are established and observed. The problem with much of the debate concerning deliberate releases has been the difficulty in getting down to specifics. Examples can be advanced that give cause for concern or that demonstrate that introductions can be carried out safely, but none of these has the generality to apply to all cases. Generic arguments for and against the safety of introductions must be rejected and replaced by consideration of the properties of individual introductions. It is the properties of the introduced organism in relation to the environment that must receive attention, not the method by which the genetic modification was achieved. One must go beyond discussions that lump all possible applications together and develop criteria that associate individual cases with the risk categories most appropriate to them.

The recent report by the US National Academy of Sciences (1987) recognized explicitly 'that the establishment of many species ...is unpredictable, and depends on the confluence of such factors as favorable weather, favorable sites, and suitable vectors or other means of transport.' The expertise of ecologists and agricultural scientists will be fundamental in reducing the unpredictability associated with introductions and in assuring that appropriate safeguards are in place. The possibilities are enormous for partnerships among genetic engineers, ecologists, microbiologists, and agricultural scientists, not only in risk assessment but also in the use of genetic engineering as a tool for ecological and evolutionary research, and in environmental management. Such cooperation is long overdue, and can ensure both that innovations are not prevented by overregulation and that legitimate concerns are not submerged by overenthusiasm.

ACKNOWLEDGEMENTS

This publication is ERC-159 of the Ecosystems Research Center, and was supported in part by the US Environmental Protection Agency, Cooperative Agreement CR812685 with Cornell University. Additional support was provided by National Science Foundation grant DMS-8406472 to the author. The work and conclusions published herein represent the views of the author and do not necessarily represent the opinions, policies, or recommendations of the funding agencies.

Thanks are extended to Clifford Gabriel for valuable comments on an earlier version of this paper and Arthur Kelman for always illuminating discussions and for calling my attention to the Thompson paper.

REFERENCES

Colwell, R.K. (1988) Ecology and biotechnology: expectations and outliers. In: Fiskel, J. and Covelo, V.T. (Eds.) *Risk Analysis Approaches for Environmental Releases of Genetically Engineered Organisms*, NATO Advanced Science Institute Series Vol. F, Springer-Verlag, Berlin, Heidelberg, pp. 163–80.

Elton, C.S. (1958) *The Ecology of Invasions by Animals and Plants*, Methuen, London.

Levy, S.A. (1982) Microbial resistance to antibiotics: an evolving and persistent problem. *Lancet* **1**, 83–8.

Levin, S.A. (1986) Risk assessment, risk management and biotechnology. In: Russell, M.J. (Ed.) *Proc. 1986 Washington International Conference on Biotechnology*, Center for Energy and Environmental Management, Fairfax, Virginia, pp. 231–44.

Levin, S.A. and Harwell, M.A. (1986) Environmental risks and genetically engineered organisms. In: Panem, S. (Ed.) *Biotechnology: Implications for Public Policy*, Brookings Institution, Washington, DC, pp. 56–72.

Mack, R.N. (1988) (and references therein). Temperate grasslands vulnerable to plant invasions: characterstics and consequences. In: Drake, J., diCastri, F., Groves, R., Kruger, F., Mooney, H.A., Rejamanek, M. and Williamson, M. (Eds.) *Biological Invasions: A Global Perspective*, SCOPE 37, John Wiley, Chichester.

Mooney, H.A. and Drake, J.A. (Eds.) (1986) *Ecology of Biological Invasions of North America and Hawaii*, Springer-Verlag, New York.

National Academy of Sciences (1987) *Introduction of Recombinant DNA-Engineered Organisms into the Environment: Key Issues*, Washington, DC, 24 pp.

Regal, P. (1988) The adaptive potential of genetically engineered organisms in nature. In: *Special Combined Issue of Trends in Ecology and Evolution* **3**(4) and *Trends in Biotechnology* **6**(4), Elsevier, Cambridge, England, pp. 536–538.

Selander, R.K., Caugant, D.A. and Whittam, T.S. (1987) Genetic structure and variation in natural populations in *Escherichia coli*. In: Neidhardt, F.C., Ingraham, J.L., Low, K.B., Magasanik, B., Schaechter, M. and Umbarger, H.E. (Eds.) *Escherichia coli and Salmonella typhimurium: Cellular and Molecular Biology*, American Society for Microbiology, Washington, DC, pp. 1625–48.

Thompson, P.B. (1987) Agricultural biotechnology and the rhetoric risk: some conceptual issues. National Center for Food and Agricultural Policy Discussion Paper FAP87-01. Resources for the Future, Washington, DC.

14 Regulation of Biotechnology: A Perspective on the US 'Coordinated Framework'

DAVID T. KINGSBURY*
Assistant Director for Biological, Behavioral, and Social Sciences, National Science Foundation, Washington, DC, USA

**See p. x for present address*

In response to public concern, Congressional interest and encouragement from the business community, on 26 June 1986 the Executive Branch of the US Government published the final part of the 'Coordinated framework for the regulation of biotechnology' in the *Federal Register* (FR 51, No. 123, pp. 23302–93). This achievement resulted from more than two years of work by a large number of people representing more than 18 federal agencies and executive offices. The goal in developing the 'Coordinated framework' was to explain to the American Public that, for questions involving the products of 'biotechnology' (more specifically, organisms derived from recombinant-DNA technology), human health and the health of the environment were of paramount concern and were adequately protected. The policy guidelines are based on widely accepted scientific principles and provide a stringent, yet rational, basis for regulation.

The process of development of the Administration's policy began in April of 1984 within a Cabinet Council Working Group. The Working Group was formed in response to public and Congressional concern over the increasing rate of research on organisms destined for introduction into the environment. At the time the process began there was general acceptance of the model of regulation of non-living products; however, there was considerable concern about the impact of new organisms introduced into the environment. The US Environmental Protection Agency (EPA) had begun to take a serious look at the regulatory problems and was working on the development of a policy. It was clear that several agencies had a stake in the process and therefore interagency coordination was essential.

In a December 1984 *Federal Register* Notice, the Working Group published a proposed coordinated framework which was open for a three month public

Introduction of Genetically Modified Organisms into the Environment. Edited by Harold A. Mooney and Giorgio Bernardi

comment period. Each affected agency attempted to define what it saw as the regulatory challenges of the technology and how the agency might respond. EPA's most critical element was a definition of a 'new' organism, the key to regulation under the Toxic Substance Control Act (TSCA). The definition proposed in 1984 was 'process' based. A microorganism would be considered 'new' if significant human intervention had been used to develop it. Microorganisms developed by recombinant DNA were presumed to be new because they involved significant human intervention.

The December 1984 Notice also included a statement from the US Department of Agriculture (USDA) regarding what constituted a 'regulated article' under USDA statutes. The Working Group proposed a strong central 'Biotechnology Science Board' to oversee (and second guess) all of the Government's regulatory activity.

Those proposals are now history. In response to the December 1984 Notice the Working Group received hundreds of public comments about all aspects of the policy. While each agency was responsible for dealing with its own policy statement the Working Group served as a central policy coordinating forum. The Working Group eventually voted to eliminate the Biotechnology Science Board and to establish the Biotechnology Science Coordinating Committee (the BSCC) in its place. The BSCC consists of the Commissioner of the Food and Drug Administration, the Director of the National Institutes of Health, the Assistant Secretary of Agriculture for Marketing and Inspection Services, the Assistant Secretary of Agriculture for Science and Education, the Assistant Administrator of EPA for Pesticides and Toxic Substances, the Assistant Administrator of EPA for Research and Development, and the Assistant Director of the National Science Foundation for Biological, Behavioral, and Social Sciences.

The Committee operates as a *scientific* coordinating committee among the agencies, within the context of its charter and a Memorandum of Understanding signed by each of the members, and is not a committee focusing on regulatory oversight. It is important to recognize two factors about the BSCC. First, its responsibility is to focus on scientific and not policy questions, and, second, it is to act as the vehicle to coordinate agency interaction and help establish a rational basis for establishing lead agency responsibility in those areas where possible overlap of jurisdiction exists.

A central question in the discussion of regulatory oversight and review of the products derived from recombinant DNA revolves around what products constitute a potential hazard, based on a sound analysis of their unique genetic makeup. To that end each of the regulatory agencies, their advisory committees, and the BSCC focused on the definitions of what should be the conditons that trigger a regulatory review. The result was to refine the process-based definition from 'human intervention' to include organisms that would *not* be 'naturally occurring' or are 'pathogens.' It was felt that microorganisms resulting from the combination of genetic material among source organisms from different genera should be considered 'new.' They were new because of the degree of human intervention required, the significant likelihood of creating new combinations of traits, and the greater uncertainty regarding potential risks of such organisms. Thus, an intergeneric organism would be a 'new' organism. Other organisms in the same taxonomic genus created by genetic engineering would not be considered new organisms. However, even these would be regulated if they were pathogenic, under the related definition of pathogen that EPA was developing.

The issue of definitions eventually came before the BSCC as the members represented the particular agencies developing them. The BSCC felt that uniformity across agencies, to the extent possible and appropriate, would strengthen and simplify the regulatory structure. Uniformity would also tend to discourage 'agency shopping' by the regulated community seeking to find a sympathetic regulatory agency.

At the time of publication of the 26 June 1986 Notice, it was thought that the appropriate regulatory balance necessary to deal with a new and rapidly developing industry had generally been achieved and, furthermore, that the principles established in this policy would be suitable as a model in other countries.

The 'Coordinated framework' is a broad and complex policy that explains the application of existing statutes to the regulation of recombinant DNA and outlines the approach to interagency coordination, which is so vital to this field.

Government regulation is not organized around technological processes, but rather the government tends to be structured around products, developed by various technologies for specific intended purposes. Therefore, one critical element in a coordinated regulatory framework is the common definition of the nature of the products subject to particular types of regulatory oversight.

The principal focus of the policy is environmental release of new organisms. There has been general acceptance of the regulation of non-living products of biotechnology and, in fact, the 1986 policy reiterates that the past regulatory practices will be maintained. The new policy explained the application of certain statutes over genetically modified organisms, and in some cases even imposes an abbreviated review over unmodified organisms when applied to environmental uses. There is a clear policy established requiring review of genetically engineered microorganisms prior to release into the environment with some organisms subject to an abbreviated review. In the unlikely event of a problem arising in this period of time EPA could use its authority under section 7 of TSCA to immediately limit or prohibit the manufacture, processing, distribution, or use of the product. In addition to the EPA activity, USDA will review all genetically engineered plant pests and animal pathogens.

The 'Coordinated framework' as a document and a policy is only the beginning. Much work remains ahead in implementation. A major activity is informational and educational. The Biotechnology Science Coordinating Committee and the involved agencies have sought to provide information to the public, and agency officials have frequently testified before Congressional committees.

While it is not possible to predict precisely what impact biotechnology will have on the future of the world economy, most observers feel it will be substantial. Recent predictions state that the current situation in the United States of relatively few sales will change to several billion dollars in the 1990s and as much as 40 billion by the year 2000. The product mix by that time is expected to be pharmaceuticals, agricultural chemicals and pesticides, growth-promoting hormones for animals and plants, and a line of very high value added industrial chemicals. Many observers predict that the use of specially engineered microorganisms to degrade pollutants and chemicals and biological waste-treatment processes are also highly probably products for the relatively near future.

It appears, to the concern of many observers, that two non-scientific issues may play a dominant role in the future of biotechnology development and implementation in the United States. First, the regulatory climate could, if not rational, present industry with an insurmountable problem for the eventual introduction of products being developed now, leading to a future withdrawal of activities in these areas. Second, concerns within the financial community about the long-term stability of companies doing business in environmentally regulated fields will lead to decreased company values and lack of capital for the continued development and testing necessary to overcome the regulatory barriers of safety and effectiveness.

Everyone associated with the regulatory questions related to the issues of biotechnology have a major responsibility to balance the various opposing forces facing this technology. We cannot accept the notion that all biotechnology-derived products are too dangerous to introduce into the environment, yet we must acknowledge the legitimate concerns of the public and work to establish those principles that govern the safe environmental use of these products and allow the underlying research to proceed. In such a fast-moving technological environment it is necessary to regularly review the appropriateness of the scientific basis of existing regulation and to make any required adjustments in either the technology of regulation or the statutory basis for regulation. The interagency committees that worked on the current policy consider the existing regulations to cover the current and near-term biotechnology industry. As the technology changes we must continue to monitor those changes and work together to keep our regulatory 'coordinated framework' modern and effective.

From a scientific viewpoint there are a number of questions that must be addressed as we view the appropriateness of various environmental applications of genetically modified organisms. These questions center on the uniqueness of the organisms and the properties that they possess that might pose an environmental hazard. The following questions are the most significant:

1. Is it inherently dangerous to use recombinant-DNA techniques to move genes between unrelated organisms?
2. Are recombinant-DNA engineered organisms like nonnative organisms?
3. Will the use of recombinant-DNA techniques accidentally create new plant pests?
4. Can recombinant-DNA accidentally convert a nonpathogen to a pathogen?
5. Can introduced genes spread in a microbial or plant population?
6. Will recombinant-DNA engineered microorganisms alter soil microbial communities?

This set of questions can be easily addressed within the context of present knowledge based on past work in a number of fields. We cannot ignore the work of the last century, especially in the agricultural areas where the breeding of new plants and development of new microbes has been the mainstay of agricultural research. If one carefully examines the above questions and develops answers based on scientific evidence the conclusion is that the potential hazards posed by recombinant-DNA technology may be identified with little difficulty and may be managed easily at the early research stage. The conclusion that may be reached from this analysis is that the role of the regulatory bodies is to examine the evidence provided by the producers and determine the veracity of the data. More importantly, however, the proper analysis leads one to conclude that the potential hazards may be much easier to identify than is frequently stated and that over time broader categories of exclusions should be developed.

15 Current Trends in the Evaluation of the Impact of Deliberate Release of Microorganisms in the Environment: A Case Study with a Bioinsecticidal Bacterium

ROBERT M. FAUST
USDA, Agriculture Research Service, Beltsville, Maryland, USA
and
KUNTHALA JAYARAMAN
Center for Biotechnology, Anna University, Madras, India

15.1 INTRODUCTION

In the past decade scientific and public debate over the use of gene splicing ultimately led to the development of guidelines and methodology to control and assess potential biohazards. Over the past several years, the attention has focused on assessment of the deliberate release of genetically modified organisms into the environment with small-scale experimental field testing viewed as a necessary part of the process in order to provide useful baseline data. Among the issues relevant to environmental assessment of the impact of the deliberate release of genetically modified microbes, plants, and animals is the question of the organism's survival, growth and multiplication, safety, and possible dissemination of its genetic information. These issues are currently being addressed by the scientific community, the various regulatory agencies, and environmental groups as deliberate release experiments are undertaken or planned for the future.

The genetically engineered microbes seem to generate the most concern because they are more difficult to monitor and the chance of horizontal transfer of genetic information is greater. Nonetheless, the vast amount of scientific information that has been generated over the years has demonstrated their potential for applications to benefit mankind. Past experience with a

Introduction of Genetically Modified Organisms into the Environment. Edited by Harold A. Mooney and Giorgio Bernardi

number of microbial pest control agents suggests that in many cases the introduction of genetically manipulated insecticidal microbes probably will be without consequence to the environment.

Numerous genetic-engineering studies of the entomocidal bacteria are already under way. Both *Bacillus thuringiensis* and *B. sphaericus* are well recognized as having the greatest potential for use in insect population control and they have received considerable attention by commercial interests. Both bacteria are characterized by the presence of a parasporal crystal composed of polypeptides. The insecticidal crystal proteins (ICP) of *B. thuringiensil* are toxic to many lepidoptera (caterpillars), diptera (mosquitoes and blackflies), and coleoptera (beetles) larvae, depending upon the strain, while the toxins of *B. sphaericus* are limited in range to mosquito larvae (Aronson *et al.*, 1986; Andrews *et al.*, 1987).

The genetics of *B. thuringiensis* has been advanced considerably over the past 10 years such that a number of important factors relevant to its use in genetic-engineering studies are now well documented in the literature. These advances may be summarized as follows: (1) the ICP genes are carried on large transmissible plasmids, (2) the ability to produce insecticidal crystals can be transferred to other bacilli by transconjugation, (3) the ICP genes have been cloned from isolates toxic to lepidopteran, dipteran, and coleopteran insects and expressed in other bacteria, (4) nucleotide and deduced amino acid sequences and protoxin subdomains are known and have been compared between subspecies of lepidopteran-, dipteran-, and coleopteran-killing subspecies of *B. thuringiensis*, and (5) possible regulatory mechanisms of ICP gene expression have been described (Andrews *et al.*, 1987; Dean, 1984; Carlton and Gonzalez, 1985; Hofte *et al.*, 1987; Sekar *et al.*, 1987; Herrnstadt *et al.*, 1987; Whiteley *et al.*, 1984, 1985, 1987; Whiteley and Schnept, 1986; Aronson *et al.*, 1986; Ellar *et al.*, 1985). The ICP genes of *B. sphaericus* have also been cloned and sequenced (Ganesan *et al.*, 1983; Hindley and Berry, 1987; Baumann *et al.*, 1987).

The highly selective action of these two entomocidal bacteria offers an environmentally safe method of insect control. *Bacillus thuringiensis* is one of several microbial control agents that has enjoyed registration with the Environmental Protection Agency (EPA). This agency is now issuing guidance to applicants on developing the required data to reflect advances in risk assessment techniques for genetically engineered microbial pesticides. The EPA is responsible for determining whether the pesticide will cause, or significantly increase the risk of, unreasonable adverse effects to nontarget organisms or the environment.

15.2 ADVANCES IN THE GENETIC MANIPULATION OF THE INSECTICIDAL CRYSTAL PROTEIN GENES OF *BACILLUS THURINGIENSIS* AND *BACILLUS SPHAERICUS*

All of the information gained from studies on the molecular biology and genetics of *B. thuringiensis* and *B. sphaericus* over the past 10–15 years has led to attempts, and undoubtedly will lead to future attempts, at providing more effective insect pest control agents based on *B. thuringiensis* and *B. sphaericus* ICPs for release into the field. Either natural gene transfer among the strains of *B. thuringiensis* to yield broader spectrum isolates (Battisti *et al.*, 1985; Gonzalez *et al.*, 1982; Klier *et al.*, 1983) or genetic engineering of the ICP gene itself are available techniques to achieve such a goal (Andrews *et al.*, 1987). Reports have begun to appear of transferring truncated or full-length sequences of ICP genes into crops such as tobacco, tomatoes, cabbage, cotton (Adang *et al.*, 1987; Andrews *et al.*, 1987; Barton *et al.*, 1987; Fischkoff *et al.*, 1987; Anonymous, 1988a) and into other bacteria associated with crop plants (e.g. *Pseudomonas fluorescens* and *Clavibacter xyli* subsp. *cynodontis*; see Crop Genetics International, 1987; de Marsac *et al.*, 1987; Obukowicz *et al.*, 1986a, 1986b; Obukowicz, 1987; Watrud *et al.*, 1985) to give unique delivery systems that reportedly will be available longer for insect ingestion.

The reports are indicative of future strategies in directing the modification of the activity of the ICPs. Such modifications should result in the production of new forms of ICPs, a broadened host range, creation of biocides possessing activities against two or more different target insects, increased persistence, ability to recycle, increased plant tolerance and resistance to insect pests, elimination of processing, manufacturing, transportation and farmer application of toxin in some instances, and perhaps increased toxin yield. Present and future strategies include the use of transforming/transducing/mating systems, recombinant-DNA manipulation/mutagenesis, molecular cloning of toxin genes (e.g. transfer of two or more genes for distinct insecticide activity in the same host cell or transfer of the ICP gene into other prokaryotes). Transfer of the toxin gene into insect viral genomes and/or transfer of the toxin gene into the insect's food source (e.g. algae, tobacco, potato, soybean, corn, cabbage, cotton, tomato, etc.) are additional strategies. The improvement of bio-insecticides to control insect pests should further decrease the dependence on chemical insecticides.

The application of intra- and interspecies transconjugation should be useful in attempts at broadening the host-range specificity by introducing several crystal gene-containing plasmids of different origins into the same recipient strain. This achievement could lead to the development of a multipurpose biocide possessing activities against lepidopteran, dipteran, and/or coleopteran insects. Strains of *B. thuringiensis* having broader spectra have already been obtained using a transconjugation system (Battisti *et al.*, 1985; Gonzalez *et al.*, 1982; Klier *et al.*, 1983). Klier *et al.* (1983) transferred the toxin gene of *B. thuringiensis* subsp. *thuringiensis* (berliner 1715), which had been transferred into *B. subtilis*, to an acrystalliferous *B. thuringiensis* mutant of subspecies *kurstaki* and also to *B. thuringiensis* subsp. *israelensis*. In the latter transfer, the resulting transcipient strain produced both types of insecticidal protein toxins and was active against both lepidopteran and dipteran larvae. Gonzalez *et al.* (1982) have also demonstrated the production of more than one protein toxin in the same cell from two different subspecies of *B. thuringiensis* (*thuringiensis* and *kurstaki*) that are each toxic to specific lepidopteran larvae.

If expression problems can be overcome, and in some cases they have been solved, the genes for the *B. thuringiensis* protein toxin might be introduced into a variety of atypical systems to improve the total efficiency of this biocide in the field. A few such efforts have recently been publicized (Crop Genetics International, 1987; Fischkoff *et al.*, 1987; Graf, 1985; Kronstad and Whiteley, 1986; Obukowicz *et al.*, 1986a, 1986b; Obukowicz, 1987; Oeda *et al.*, 1987; Prefontaine *et al.*, 1987; Watrud *et al.*, 1985; Anonymous, 1988a).

For example, the gene might be introduced into bacteria commensurable with plants. Seeds could be watered with these modified bacteria, and the seeds and resultant seedlings may be afforded some measure of protection against susceptible insect pests as well as the growing plant. Both the Monsanto Company and Mycogen have cloned the protein gene of *B. thuringiensis* subsp. *kurstaki* into *Pseudomonas* species. The Monsanto Company's biocide, a recombinant *P. fluorescens*, which in the natural state is presumably found on the roots of corn plants, is intended to colonize corn roots for control of the black cutworm, *Agrotis ipsilon*. Corn seeds can be coated with the preparation. To ensure that the toxin gene is not transferred to another microbial species, the company's scientists inactivated the transposase enzyme to prevent the movement of the transposon, which is associated with the toxin gene. Mycogen's approach involved creating a 'biological package', or microcapsule, for the toxin to protect it from environmental stresses. The company 'fixed' or killed the *Pseudomonas* organism for use in the field. The killed vegetative cells presumably protect the toxin, thus preventing premature deactivation. Mycogen suggests that using killed cells should make the product easier to formulate and handle. More recently, Crop Genetics International (1987) has inserted the ICP gene from *B. thuringiensis* supsp. *kurstaki* into the chromosome of the endophytic bacterium, *C. xyli* subsp. *cynodontis*, for control of the European corn borer, one of the largest uncontrolled insect pests of corn (Calvin *et al.*, 1988). The endophytic microorganism is capable of colonizing corn internally.

Other bacteria, such as natural pond microflora, might be suitable hosts for insertion of mosquitocidal crystal protein genes and field release. The suitability of the cyanobacteria for molecular cloning purposes has improved over the past 5 years with the identification and construction of useful shuttle-cloning vectors. These cloning systems in which the mosquitocidal protein gene from *B. thuringiensis* could be introduced into a blue-green algae and maintained on an autonomous replicon should be helpful in developing algae-based mosquito controls having persistance and ability to recycle.

Plant Genetic Systems scientists have succeeded in inserting the mosquitocidal gene of *B. thuringiensis* subsp. *israelensis* into a blue-green algae (personal communication). Additionally, de Marsac *et al.* (1987) have reported transferring the mosquitocidal gene of *B. sphaericus* into the cyanobacterium, *Anacystis nidulans*. The fragment was initially cloned and expressed in *E. coli* and *B. subtilis* using pHV33 as a shuttle vector, and then cloned into the blue-green algae with a pUC303 shuttle vector. The cyanobacteria grow on the upper layers of aquatic habitats, persist relatively long in the environment, and presumably should be more effective in controlling mosquito larvae. These reports are good indicators of potential development of biocides with amplified persistance in mosquito habitats.

Other recent efforts at developing novel biocides involve incorporating the toxin genes directly into the genetic background of plants, thereby allowing the plant to synthesize its own insecticidal material. The success of such efforts should result in bypassing the need for repeated applications of the agent, and eliminate the processing, manufacturing, and transportation of insecticidal products. Such modifications have been made possible since both efficient DNA vector constructs are now available for use in transporting foreign genes into higher plant cells. *In vitro* systems for genetic engineering of plants have also been developed for a number of crop plants including corn, cotton, soybean, cabbage, tobacco, tomato, potato, lettuce, carrot, sugar beet, etc.

The vectors for transferring foreign genes such as the ICP gene into higher plant cells or tissues currently are based on *Agrobacterium tumefaciens* or the cauliflower mosaic virus, although other sources of vector materials being studied include *A. rhizogenes*, other DNA caulimoviruses, geminiviruses (e.g. bean golden mosaic virus), potato leafroll virus, and liposomes (Andrews *et al.*, 1987). The Rohm and Haas Company (plant genetic systems), Agrigenetics Advanced Science Company, Agricetus, and the Monsanto Company have reported that the *B. thuringiensis* toxin gene of subsp. *kurstaki* has been inserted into tobacco where it is expressed (Freedman, 1985; Adang, 1987; Andrews *et al.*, 1987). Both leaves and the tobacco plant tissues synthesize the insecticidal toxin. Barton *et al.* (1987) have also reported successful transfer and expression of the *B. thuringiensis* ICP gene in tobacco where it provides the plants with resistance to lepidopteran pests.

Tobacco plants carrying the toxin gene should be resistant to tobacco budworm, cabbage looper, and the tobacco hornworm.

Other important agricultural insects susceptible to the *B. thuringiensis* subsp. *kurstaki* toxin are the corn earworm, cotton bollworm, and the beet armyworm. Agracetus has reportedly developed cotton plants that harbor the ICP gene and are resistant to the major lepidopteran pests of this important crop. Under laboratory conditions, the transgenic cotton plants deter feeding and cause death of the pest larvae that do feed. Subsequent greenhouse testing confirmed the resistance and also demonstrated that the ICP gene was passed on to the seed and subsequent generations (Anonymous, 1988a). A few other publicized reports indicate that work is underway on transferring the toxin gene to cabbage and potato.

A significant step in developing insect-tolerant transgenic plants other than tobacco and cotton was recently reported by Fischkoff *et al.* (1987). Two truncated genes from *B. thuringiensis* subsp. *kurstaki* were incorporated into a plant expression vector for *Agrobacterium*-mediated transformation. The genes were transferred into tomato plants where they were expressed. Expression of the ICP genes conferred insect tolerance on the transgenic tomato plants and their progeny. Larvae (*Manduca sexta, Heliothis zea*, and *H. virescens*) were killed within 48 hours with very little evidence of feeding damage to the leaf. Some of the transgenic plants were able to kill 100% of the *M. sexta* and *H. virescens* larvae. The investigators also obtained preliminary evidence that some toxin activity was detectable in tomato fruit.

The future development and field use of such novel microbial insecticide products by genetic manipulation will require the availability of efficient and appropriate cloning/transfer/promoter systems that incorporate the technical advantage of high-level expression of a variety of *B. thuringiensis* ICP genes. Transfer of important genes to select procaryotes and to higher plants appears to be a facet of intense research that is now beginning to yield significant results. Application to the regulatory agencies for environmental release permits have concomitantly increased.

15.3 DELIBERATE RELEASE AND ENVIRONMENTAL ASSESSMENT OF A BIO-INSECTICIDAL BACTERIUM—A CASE STUDY

The Environmental Protection Agency (EPA), the United States Department of Agriculture (USDA), and the Food and Drug Administration (FDA) have established protocols and procedures for implementing their policies for prerelease reviews of proposals involving genetically engineered organisms. These agencies have identified general data requirements and possess the authority to request additional data when necessary to evaluate individual proposals. The USDA and EPA may combine the wide-ranging expertise of the various federal agencies in their reviews and also coordinate with state regulatory officials. The Federal Insecticide, Fungicide and Rodenticide Act (FIFRA) prohibits the distribution, sale, and use of pesticides that have not been registered with EPA, including microbial pesticides. The EPA must review all data submitted on each microbial pesticide before determining whether it should be registered. The Agency has issued guidelines to applicants on developing the required data and continues to revise the guidelines to reflect advances in risk-assessment techniques for genetically engineered microorganisms.

In order to gather product performance and other data necessary for the application, producers may obtain an Experimental Use Permit (EUP) to conduct field studies. The EPA and USDA have indicated that as they gain knowledge and experience from case-by-case reviews, they expect to develop broadly applicable procedures and guidelines. The agencies contend that, for the most part, existing laws available for the regulation of products developed by traditional techniques should be adequate for regulating genetically engineered organisms. Also, all federal agencies are required under the National Environmental Policy Act (NEPA) of 1969 to prepare an analysis before taking a major action that may significantly affect the environment. The agencies first perform a preliminary assessment of the possible consequences of an action to determine whether to prepare an Environmental Impact Statement (EIS) or a Finding Of No Significant Impact (FONSI). If the Environmental Assessment (EA) indicates a significant environmental impact, the Agency must prepare a detailed EIS. The EPA strategy is to regulate genetically engineered microbial products using existing authority with some additional rulemaking for its chemical control statute (Anonymous, 1988b). Unlike the USDA and the EPA, the FDA at this writing does not plan to promulgate new rules for regulation.

In December of 1987 Crop Genetics International (CGI) of Hanover, Maryland, transmitted an application to the EPA and to the USDA for approval to field test a recombinant endophytic bacterium, *Clavibacter xyli* subspecies *cynodontis* (Cxc), that is capable of colonizing the vascular system of corn. The organism was engineered to harbor the ICP gene of *B. thuringiensis* subsp. *kurstaki* HD73 (Btk) and express the ICP active against the larval stages of the European corn borer (ECB), *Ostrinia nubilalis* (Hubner). Field testing was planned for May to October of 1988, at the CGI Research Farm, Ingleside, Maryland, and at the USDA Agriculture Research Center, East Farm, Beltsville, Maryland. Additionally, CGI planned to conduct a field test of the Cxc/Btk recombinant bacterium with the Institut National de la Recherche Agronomique (INRA) at Montfavet, France, in 1988.

This field release has been selected as the case study because CGI not only chose to claim limited confidentiality but also chose to leave essentially all of their submittal open to the public. Additionally, field releases of genetically engineered bacteria seem to generate the most concern and represent rather unique monitoring issues. This particular field test was designed to ascertain specific information on the fate of a genetically designed bio-insecticidal bacterium in the environment.

The main objectives of the field study was to ascertain the effectiveness of Cxc/Btk (InCide) as a corn associated insecticidal agent for control of the ECB and to obtain further knowledge on the behavior of the engineered microbe in the environment. The EUP application included data-addressing environmental persistence and propagation, genetic stability, effects on nontarget plants, and toxicology (human health effects). The Hazard Evaluation Division (HED) of the EPA presented their final scientific position on CGI's EUP application for the field test on 21 May 1988 (a preliminary scientific position was completed on 25 March 1988, forwarded to an EPA workgroup, other Federal agencies—NIH, NSF, FDA, USDA—and made available for public comment).

The Animal and Plant Health Inspection Service (APHIS) reviewed CGI's permit application for a permit for release to the environment of a regulated article, and on 20 May 1988 a document giving notice that the USDA intended to issue a permit for release into the environment of a regulated article under regulations issued pursuant to the Federal Plant Pest Act and the Plant Quarantine Act was signed. The document also contained an EA and FONSI. After reviewing the application, APHIS concluded that the proposed field test of the genetically engineered microorganism did not present a risk of introduction or dissemination of a new plant pest and would not have a significant impact on the quality of the human environment. Several supplemental permit conditions, as well as the standard permit conditions, were also included. The EA and FONSI were prepared in accordance with NEPA, regulations of the Council on Environmental Quality for implementing the procedural provisions of NEPA, USDA regulations for implementing NEPA, and APHIS Guidelines for implementing NEPA.

Prior to the approvals of EPA and APHIS the Institution Biosafety Committee (IBC) at the Beltsville Agriculture Research Center reviewed the proposed field studies and agreed that the field study was of low risk and was not likely to exert an adverse effect on the environment; nor was the genetically engineered organism likely to be lethal to humans or to nontarget organisms, provided that the study was conducted in accordance with the containment, monitoring, and contingency procedures and modifications detailed in the documents submitted to the IBC. The IBC gave its approval on 12 May 1988 contingent on the issuance of an EUP from the EPA under regulations issued pursuant to FIFRA, the issuance of a permit for introduction of a genetically engineered organism from APHIS under regulations issued pursuant to the Federal Plant Pest Act (FPPA) and the Plant Quarantine Act (PQA), and fulfillment of any requirements of NEPA.

The findings and approvals for the field tests were based on CGI's scientific data which was submitted in six volumes of documentation and on in-depth discussions, formal reviews, and miscellaneous other documentation requested by the reviewing organizations. Materials in the volumes provided background information on the wild-type endophytic microorganism and compared the biological characteristics of the endophyte and the recombinant-DNA organism, descriptions of the techniques required to achieve transformation, a description of the ICP gene and the integration vectors used to insert the desired genetic material, including antibiotic resistance genes, a description of the resulting recombinant constructions containing the ICP gene, extensive information addressing the toxicology of the genetically engineered organism, and a description of the proposed field tests and the experiments to be conducted with the recombinant-DNA organism.

Initially, CGI requested that APHIS concur with a contention that *C. xyli* subsp. *cynodontis* was not a regulated article; however, APHIS, after a search of the scientific literature, concluded that the organism was capable of inciting 'disease or damage' to bermudagrass, and therefore was a 'regulated article' under the provisions for issuing a permit for the introduction of an article regulated by them. Hence, this is the reason why CGI had to receive a permit from both agencies before conducting the field test. The 'Coordinated Framework for Regulation of Biotechnology,' published in the *Federal Register* on 26 June 1986, established a structure for Federal biotechnology regulation and provided for coordination between agencies when jurisdiction is shared. The 'Coordinated Framework' provided that, when an application is for release into the environment of a microorganism intended to act as a pesticide subject to the provisions of FIFRA, EPA is to serve as the lead agency. In the review process APHIS and EPA fully coordinated their separate reviews under their respective authorities.

This particular case study represents one example of what is involved in the preparation for, and evaluation of, deliberate release of a genetically engineered bacterium. Most of the following discussion of this example is excerpted from the EA document and the materials submitted by CGI to EPA, APHIS, and the IBC (BARC Biosafety Committee, 1988). The major source of the discussion, however, is exerpted from the APHIS document (USDA, APHIS, Washington, DC).

The finding by APHIS that there would be no significant impact on the environment from the field test was based on a number of factors documented by CGI in their submitted materials and summarized by APHIS as follows:

1. The inserted ICP gene is lost from Cxc at the rate of once in every 1.25×10^4 bacterial cells per generation, and both the revertant (which has lost the ICP gene) and the naturally occurring Cxc grow faster (14 percent growth rate differential in laboratory medium) than the recombinant bacterium (in fact, it was shown that the ICP gene will

eventually be lost from the recombinant bacterium).

2. The genetic alterations are not expected to enhance any plant pathogenic property of the recombinant organism compared to the parental strain of Cxc that occurs naturally in Maryland where the test was to take place.
3. Although Cxc can under limited circumstances be transferred by mechanical means (e.g. cutting tools) to other plants, it is not transferred easily by other mechanisms in the field. Transfer to other plants by mechanical transfer is minimized by the field test design and field test protocol which include buffer zones and tool disinfestation. In addition, regular monitoring for the recombinant bacterium will ensure that if it spreads to plants at the edge of the test plot it will be detected.
4. Dissemination of Cxc can occur in seed, so all seed not used for research purposes (in containment) will be destroyed, preventing transfer by this mechanism.
5. Data were provided by CGI to demonstrate that the probability of transfer of the ICP gene from the recombinant bacterium to other microorganisms is extremely remote.
6. The recombinant bacterium has a relatively low order of toxicity to susceptible insects and since the field test plots are very small, the introduction of the recombinant bacterium would pose no significant impact on insect populations.
7. No threatened or endangered insect species are present in Maryland, so the introduction of the recombinant bacterium would pose no threat to these insects.
8. The inherent properties of Cxc and the recombinant bacterium indicated that there would be no human health risks since the organisms do not grow at human body temperature and they have been demonstrated not to be pathogenic or toxic in mammalian tests. In addition, all crops would be used for research purposes or destroyed so there would be no dietary exposure to humans.

Additionally, the recombinant bacterium was tested in the greenhouse to obtain initial data relating to the biology of interaction of the bacterium with plants and insects and to obtain preliminary data on the interaction of the bacterium with plants and insects and to obtain preliminary data on the interaction of the recombinant bacterium with plants and its efficacy in controlling insects. APHIS noted that it was normal for controlled field tests to be performed after greenhouse testing to confirm the efficacy data and to collect additional data on the complex interactions of organism and that such tests can only be validated in the environment using standard agricultural practices; such limited field testing is essential for the development of a potential agricultural product.

In its evaluation APHIS examined the proposed field plot design, field test protocol, and other factors necessary to identify the aspects of the environment that would potentially be affected. The agency also examined in detail the biology of the recipient/donor organisms involved in constructing the recombinant bacterium and attempted to identify the potential impacts to the environment inherent in those components and describe the ways in which the risk to the environment is limited, either by the nature of the organism or by specific safeguards that had been designed into the protocol.

For their studies CGI developed a number of methods to detect and monitor Cxc and Cxc/Btk organisms in soil, water, plants, and plant debris. Semi-quantitative population estimates of the organisms in extracted or expressed plant sap were performed by phase-contrast microscopy with a sensitivity of 1×10^6 cells/ml of plant sap. A more sensitive assay used, quantitative isolation using a plating technique, had a detection limit of 2×10^2 CFU/g of fresh weight of colonized plant tissue. Spontaneous antibiotic resistant mutants of Cxc were developed for monitoring the organism's persistance in the environment. One Cxc strain was resistant to rifampicin and streptomycin, while a second strain was resistant to rifampicin, streptomycin, and spectinomycin. Using selective media, the detection limit for the doubly resistant strain in corn field runoff water was 1.2×10^2 CFU/ml. For the triply resistant strain the detection limits were 4.3×10^1 in river water and 3.4×10^4 in field soil. Detection levels for the Cxc/Btk recombinant bacterium were similar to the spontaneous antibiotic resistant strains (2.7×10^1 CFU/ml, non-sterile water; 2.6×10^4 CFU/g, non-sterile soil).

A radioimmunoassay was also developed for the detection of Cxc (no cross-reactivity was observed with any other coryneforms, except *C. xyli* subsp. *xyli*, or other plant-associated bacteria tested). The detection limit for the assay was 1×10^6 cells/ml of extracted plant sap.

The data submitted by CGI demonstrated that bermudagrass colonized by the naturally occurring strain of Cxc may be stunted under growth-limiting conditions, such as light limitation and high humidity; however, the stunting was reversed when colonized plants were returned to normal growing conditions. In addition, CGI demonstrated that the yield of corn colonized by the naturally occurring strain of Cxc after artificial inoculation may be reduced in some cases by 2–17%, depending on the corn strain and geographic location. APHIS concluded that these observed effects on corn would not be a problem for the field test since Cxc occurs naturally in Maryland, and no reports of it affecting corn yield have surfaced. APHIS also concluded that the field test would not appreciably increase the total number of Cxc or Cxc/Btk organisms in the environment; nor would the engineered strain effectively compete with the parent organism in terms of growth rate.

Since there were relatively few literature references on Cxc, CGI generated extensive data on its environmental traits. These data included studies of the growth and survival of Cxc in soil, water, and plants, surveys of the natural occurrence of Cxc, and dissemination studies designed to determine if, and how, Cxc may be transmitted to other plants. (APHIS noted that environmental rate studies are inherently limited by the sensitivity of the methods used to detect the bacteria and that there are no detection methods that can reliably detect very low levels—approximately 10 bacteria/gram of soil—of bacteria in environmental samples.) Under laboratory conditions, Cxc was found to colonize a wide range of plants when those plants were inoculated and then screened by the expressed sap method. Colonization was detected in 83 species of 26 families, and no colonization was detected in 52 species of 26 families. Some families included both host and nonhost species. These plants were inoculated with a relatively high level of Cxc (approximately 10^{10} CFUs/plant) through a puncture wound using a scalpel blade. Although many plants were shown to support growth of Cxc, no adverse effects were observed despite the presence of greater than 10^6 cells of Cxc/ml of sap in the colonized plant.

In order to evaluate actual dissemination rates in the field, CGI sampled corn plants that were adjacent to corn plants experimentally colonized with Cxc. APHIS noted that the results were inconclusive (3 of 2264 sampled corn plants were colonized), since a relatively insensitive detection method was used, and the background rate of infection was not determined. The three colonized plants were found within a few meters of the inoculated test plants. No Cxc was found in 32 bermudagrass plants or 5 bindweed plants that were grown in close association with the intentionally infected corn plants.

In addition to the field surveys, CGI submitted studies designed to examine how Cxc could be transmitted in the soil. In greenhouse studies corn was planted in soil containing high levels of Cxc; no colonization of the corn was observed even after deliberate root wounding. Nor was Cxc detected in corn grown in field plots that previously harbored Cxc-infected corn plants tilled into the soil. CGI also submitted data demonstrating that Cxc can not be passed through corn seed to its progeny and thus this mode of transmission was not considered an important factor by APHIS in the field test since the seed would be destroyed. The potential for insect transmission of Cxc was tested in three species of insects (the green sharpshooter, *Draeculacephala minerva*, the red-headed sharpshooter, *Carneocephala fulfida*, and the armyworm, *Pseudaletia unipuncta*) with negative results.

The environmental fate studies consisted of laboratory studies, field surveys, and field transmission studies which APHIS noted was limited by the sensitivity of the detection methodology. Although some of the laboratory data on the growth and survival of Cxc in soil, water, and plants were limited by the use of relatively insensitive methods to screen for the presence of the bacteria, APHIS felt that the data generally supported the conclusions that Cxc does not survive well in soil and water, but grows well in plants.

APHIS felt that it was not possible to conclude with confidence that the field survey data provided by CGI on the lack of plant colonization was valid, since the rate of low-level colonization could not be ascertained. In the studies, either the numbers of plants sampled were lower than the ratio of positive to negative plants in the corn-dissemination experiment or relatively insensitive screening detection methods were used. The corn transmission studies also used less sensitive detection methods. Nevertheless, in many cases, APHIS felt that these methods were useful in providing a general indication for the potential for colonization since CGI showed, using their most sensitive detection methods (10^2–10^3 CFUs/ml), that when young susceptible plants were inoculated with Cxc, they invariably supported growth of Cxc to levels detectable by the less sensitive screening methods developed by CGI, indicating that the plants function as an enrichment mechanism for infective doses of Cxc in the environment. By these methods, CGI was also able to detect colonization in the laboratory with inoculated plants used for the host range survey. APHIS noted that these considerations allow for use of these studies as a general indication of the potential for environmental colonization.

The data indicated that Cxc can infect many plants after direct inoculation (an infective dose for corn was demonstrated at 100 CFUs/plant), but did not appear to naturally inhabit certain susceptible plants surveyed by CGI in the field. Cxc was shown to occur naturally only in bermudagrass, and the incidence of transmission from artificially colonized corn plants seems to be very low.

CGI demonstrated that there was no gross colonization of sugarcane from infected bermudagrass and no gross colonization of bermudagrass or bindweeds in fields of corn colonized by Cxc. APHIS concluded that Cxc is not likely to persist at levels high enough to lead to plant infection in water or soil, except at temperatures too low for plants to grow. This conclusion was supported by studies showing that corn was not infected by growing in soil infested with high levels of Cxc either in the laboratory or in the field. The data indicated that there may be an avenue for limited movement of Cxc to other plants, particularly by mechanical transfer (i.e. by cutting tools) or by transmission through seeds, but APHIS noted that the rate of movement was of a low magnitude and should be controlled with proper procedures. Thus, in the field test all tools used in the plots should be disinfested with a 10% solution of chlorine bleach. The seed produced in the plots should be destroyed or, if not, used in carefully controlled scientific experiments.

CGI submitted data demonstrating that the probability of gene transfer to another organism was extremely remote. No plasmids or prophage were found to be harbored by Cxc. No conjugation occurred in a filter mating experiment with *C. flaccumfaciens* subsp. *cortii*, the only related *Corynebacterium* known to have a conjugative system (Vidaver and Starr, 1981). APHIS noted that CGI truncated the transacting recombinase gene segment in their gene cassette to further reduce the possibility of gene transfer. The ICP gene was integrated into the chromosome of Cxc. In fact, in a 70 generation growth experiment, CGI demonstrated that the engineered genes were lost from the parental Cxc at a segregation loss frequency of 0.6%, which was equivalent to a calculated value of 8×10^{-5}/cell per generation. APHIS pointed out that this rate was almost 10^4 greater than the natural mutation rate in bacteria, and it also is undirectional, i.e. it will result in irreversible loss of the gene as compared to mutational events which may be reversible, indicating that the ICP gene would eventually be eliminated from the population of bacteria that would develop from the recombinant bacterium introduced to the field.

APHIS additionally requested that CGI provide generation times and any supporting evidence for estimating gene loss time. The information provided showed that the calculated rate of ICP gene loss by the recombinant bacterium in corn plants was 7×10^{-5}/cell. The estimated frequency of revertant/corn plant was 0.25%. CGI also provided information that the parental Cxc was able to grow about 14% faster than the recombinant bacterium *in vitro*. Revertants likewise grew faster than the recombinant bacterium. Although the rate of ICP gene loss was not precisely known, the information that the presence of the gene decreases the growth rate of Cxc confirmed the conclusions that, in the absence of selective pressure, the recombinant bacterium would eventually be outcompeted by the faster-growing Cxc revertant bacteria that have lost the gene. Additionally, APHIS concluded that there should be no selective advantage to the presence of the gene since CGI demonstrated that Cxc/Btk could not be recovered from the European corn borer or its frass after being fed the recombinant bacterium-infected plants for 2 weeks. In APHIS's view, the target insect would not provide a reservoir for enrichment of the recombinant-bacterial population.

APHIS also addressed issues relevant to the effects on nontarget insect populations, other nontarget animals, nontarget plant communities, exposure of threatened and endangered organisms, and possible impact on the immediate physical environment and on human health from the data submitted. It was noted that CGI's Cxc/Btk strain had a relatively low order of toxicity to the target insect as compared to the parent *B. thuringiensis* subsp. *kurstaki*. In addition, the ICP gene would eventually be eliminated from the engineered population. Beneficial invertebrates, such as honey bees, would not likely be affected because the ICP is not toxic to honey bees. APHIS concluded, therefore, that the toxic hazard to nontarget insects would be minimal. APHIS also noted that, since the field plots would be surrounded with a chain-link fence, ingress to the plot by large vertebrate animals would be prevented. No factor unique to the field test was identified by APHIS that would have an effect on birds, wild animals, or vertebrate or invertebrate (except target lepidopteran insects) animals. APHIS further noted that, as described in data provided by CGI, at least 56 species of plants can be infected with Cxc when mechanically inoculated in the greenhouse with large doses of bacteria; however, with plants inoculated by CGI in their greenhouse studies, no signs of disease were detected.

APHIS concluded that dispersal of the recombinant organism from the test plot would be prevented or effectively limited by the field test design and protocols. Trap plants would be used to detect limited movement of numbers of bacteria sufficient to infect plants. Trap plants of both corn and the natural host, bermudagrass, would surround the test plots. The trap plants would be separated from the test plots by a strip of soil that is to be barren, but the plants would be located within the fence of the field plot. These trap plants would serve as a sensitive indicator not only of dispersal by known mechanisms, such as mechanical dispersal by phytophagous insects, and 'of the even more remote possibility' of dispersal by windblown pollen.

Since the recipient organism is widely prevalent in Maryland and the recombinant bacterium was not altered to increase its pathogenicity to plants, APHIS concluded that the inserted gene would be eventually diluted or lost, and dispersal of the recombinant organism from the field test plot would be effectively limited with no risk to nontarget plants. Since the field tests would take place in the State of Maryland, while endangered lepidopteran species are in Washington, Oregon, California, and Florida, and since the dissemination from the test site and toxicity to insects would be minimal, APHIS also concluded that use of the recombinant bacterium under the conditions of the field test would not pose any risk to endangered/threatened species.

As to a possible impact on the immediate physical environment, APHIS noted upon examination of the CGI data that when levels of Cxc were inoculated into soil; the populations declined to undetectable levels (10^4 CFUs/gm soil) after 1–28 days depending on the soil conditions at temperatures from 4 to 50°C. CGI demonstrated that stable levels of Cxc could be recovered up to 28 days when kept in soil at −5°. Similar results were observed in water except that extended survival was seen at +5 and −5°C. APHIS noted that although it is not possible to show that a microorganism is completely eliminated from the environment using this type of experiment, the data demonstrated that Cxc did not grow well in soil and water, but could survive at low temperatures. No potential, however, was seen for population levels of Cxc to increase in soil or water.

Because Cxc is a natural bacterial resident of plants and owing to its somewhat fastidious growth requirements, APHIS concluded that it was extremely unlikely that Cxc could infect mammals. The conclusion by APHIS of a lack of human health risk was further supported by the lack of pathogenicity or toxicity in CGI studies of the Cxc/Btk in laboratory mice. In these studies test mice were dosed by a number of exposure routes (oral, pulmonary, and intravenous) with a high titer (about 10^8 bacteria) of the recombinant bacterium; no toxicity, infectivity, or pathogenicity was observed. Furthermore, in answer to an APHIS request for additional information, CGI presented data that the parental strain (*C. xyli* subsp. *cynodontis* MD69a) grows slowly at 34°C and does not grow at 36°C or normal human body temperature. Additionally, APHIS pointed out there has been no confirmed records of human health hazards associated with exposure to *B. thuringiensis* var. *kurstaki*, which has been widely used for more than 20 years to control susceptible insects. Extensive pathogenicity/toxicity testing with *B. thuringiensis* in laboratory animals and in human volunteers has supported the lack of health risks associated with exposure to this bacterium and to the ICP.

The two test sites in the United States were of similar plot design with only minor changes in the width of the barren and fallow zones and differences in the number of subplots to accommodate the size of each site. APHIS noted that the plots were designed to allow for efficient maintenance, ease of personnel movement between individual subplots, separation of specific, short-term studies from those studies of longer duration, and the detection of any dispersal of the recombinant bacterium outside the confines of the plot. It was noted that the plots consisted of (1) central corn plant populations arranged in distinct subplots, (2) a barren plant-free zone adjacent to and surrounding the subplots, (3) a 'trap' plant zone consisting of corn and bermudagrass bordering the barren zone, (4) a black-plastic-covered earthern dike 2 feet wide and 18 inches high, constructed around the test site, just inside the security fence, (5) a standard 8 foot high, chain-link security fence, and (6) a low-maintenance fallow zone of naturally growing weeds (sampled and analyzed for the recombinant bacterium at intervals throughout the season to test for possible movement of the microbe from the test sites outside the security fences). The corn was irrigated as needed by trickle irrigation and a security guard was on location 24 hours/day.

Conditions for the test included the requirement that plants and soil should be monitored in the tests by collection of the following data: (1) plant colonization by the recombinant bacterium 4 weeks after inoculation, (2) colonization by the recombinant bacterium of all plant parts monthly for 4 months, (3) natural and mechanical dispersal of the recombinant bacterium in the field after 60 days, (4) presence of the recombinant bacterium in runoff water, (5) presence of populations of the recombinant bacterium in the soil, (6) effect of the recombinant on crop yield; (7) effect of the recombinant bacterium on crop residue decomposition, (8) effect of the recombinant bacterium on vesicular-arbuscular mycorrhizae 3 and 6 weeks after planting, and (9) effects of the recombinant bacterium on populations of saprophytic gram-negative bacteria in the phylloplane.

APHIS also felt that in the event the presence of corn flea beetles, leafhoppers, and adult southern corn rootworm were detected in association with inoculated test plants, any insects present should be analyzed for the presence of the recombinant bacterium. In all analyses that involved testing for potential dissemination of the recombinant bacterium (trap plants, indigenous plants in the fallow zone, corn residues in the field), CGI should use a detection method at least as sensitive as the plate-count method using plant homogenate (theoretical limit of detection of 2×10^2 CFUs/g). The weed species that grew in the fallow zone areas should be monitored for the presence of the recombinant bacterium to detect any potential spread from the test plot, and when surveying the potential for dispersal into naturally occurring weeds during the field test, all weed species tested should be selected from those weeds known to be susceptible to infection when artificially inoculated with the recombinant organism or parent organism. Also, prior to initiation of the field tests, a survey of indigenous plant species in the fallow zone of each test site should be carried out. All major plant species found should be tested for susceptibility to infection by artificial inoculation. APHIS stated that, during monitoring, the majority of samples from the fallow zone should be taken from plants known to be susceptible to infection.

A number of other containment protocols were discussed by the reviewing organizations, in addition to the suggested containment provisions already mentioned, and included disinfection of field equipment (tractors, rototillers, sprayers, cultivators, hand tools, etc.). Machinery used in the field test should be disinfested at the edge of the test plot just inside the barren zone, and all organic debris should be thoroughly cleaned from surfaces to be disinfested (cutting blades, tires, etc.). All surfaces should be sprayed with a 10% solution of household bleach (final concentration 0.525% sodium hypochlorite) for a period of 10 minutes in order to make sure they were continuously wet. In addition, a surfactant should be used to ensure uniform wetting of the surfaces and all disinfested surfaces should be allowed to dry before removal from the field. Small tools and equipment used for inoculation and sampling should be decontaminated on site by continuous submersion in 10% solution of household bleach for a minimum of 10 minutes.

For the transfer of the recombinant bacterium from CGI's laboratories in Hanover, Maryland, or the field laboratories, to the site of release the reviewing organizations suggested movement be carried out in secured, sealed containers. Additionally, investigators and visitors should sign in and sign out at the field test site. The sign-in/sign-out sheets should be maintained as part of the field study records. Footware should be disinfested before exiting the barren zone and all personnel involved in the field test experiments would be instructed in the environmental safety aspects and record-keeping requirements of the test.

APHIS also suggested that bermudagrass and corn trap plants be planted in the soil rather than in containers, as CGI originally specified in the application. The plants would thus be exposed to the soil and runoff water and would provide a more realistic environmental exposure. The bermudagrass should be tested and certified free of infection prior to planting. Each trap plant should be sampled monthly along with all plant parts of inoculated corn plants.

Inoculated corn plants should be surveyed weekly for the presence of flea beetles, leafhoppers, and the adult southern corn rootworm, and any collected insects should be analyzed for the presence of the recombinant bacterium. If the recombinant bacterium was detected in the insects surveyed from the infected corn plants or if the recombinant bacterium was detected in the corn trap plants, the test site should be treated with an insecticide that was effective against the insect found harboring the recombinant bacterium. APHIS had to be notified of the findings as soon as possible, but not later than 5 working days. Also, if the recombinant bacterium was detected in the trap plants (bermudagrass or corn), APHIS had to be informed as soon as possible, but not later than 5 working days. CGI, within 30 days, would take samples of the indigenous weeds in 200 one-foot square plots located in the fallow zone outside of the fence. This sampling would be conducted at least every 2 weeks following detection of the recombinant in trap plants and would be conducted as a replacement for, or in addition to, the monthly fallow zone sampling. If the recombinant bacterium was found in a sample from the fallow zone, APHIS had to be informed within 24 hours and the experiment had to be terminated.

If termination of the field test was required, a broad-spectrum herbicide should be used to kill all plant material in the test site and the fallow zone, all plant material should be collected and incinerated, the test site and the fallow zone should be fumigated with methyl bromide, and a layer of soil from elsewhere on the test farm should be spread over the test site and the fallow zone, and tilled in to provide an inoculum of naturally occurring organisms subsequent to the methyl bromide fumigation.

As specified by APHIS, a normal termination of the experiment could be accomplished by following the same procedure. Alternatively, CGI could monitor the test site in the following year (1989) by planting the site with bermudagrass and sampling the grass for the presence of the recombinant bacterium. If the recombinant bacterium was detected, then the supplemental monitoring could be terminated and the previously prescribed termination procedures should be followed. If no recombinant bacterium was detected, then fumigation and soil incorporation would not be necessary.

15.4 CONCLUSIONS

In a number of instances the potential to manipulate the genes of entomocidal bacteria has become a reality. Within the past ten years much has been learned about the genetics and molecular biology of these organisms and advances in gene manipulation and gene transfer technology have made possible the introduction and expression of foreign genes, including the *B. thuringiensis* and *B. sphaericus* entomocidal protein toxin, in transgenic organisms.

The location of the insecticidal crystal protein genes on transmissible plasmids permits construction of novel insecticidal combinations by transconjugation between diverse isolates. Through molecular cloning, gene transfer, recombinant-DNA manipulation, and site-directed mutagenesis more sophisticated biocides will undoubtedly be developed, and progress has already been made using some of these strategies. The nucleotide and deduced amino sequences and protoxin subdomains are now known for a wide variety of the *B. thuringiensis* subspecies that affect at least three orders of insects. Their comparisons have given some insight into the differences between their activities. The use of gene splicing methods and reconstruction of the various toxin domains should allow for greater insight and perhaps the creation of novel biocides based on *B. thuringiensis* (and *B. sphaericus*) that may exceed the present-day naturally occurring isolates in efficacy.

Products based on the ICP gene and developed from this technology have been appearing in increasing numbers. Controlled field releases are needed so that information on the efficacy and biology of the interaction of genetically engineered insecticidal bacteria with the environment can be obtained. In the case study summarized in this report, greenhouse tests were completed to obtain peliminary data on the interaction of the recombinant bacterium (Cxc/Btk) with the environment and the efficacy of the insecticidal organism to control insects. In an environmental assessment for the issuance of a permit by APHIS for Crop Genetics International (CGI) to conduct field tests with engineered *Clavibacter xyli* subsp. *cynodontis* harboring the insecticidal crystal protein gene of *B. thuringiensis* subsp. *kurstaki*, the APHIS stance was that it is normal for controlled field tests to be performed after greenhouse testing to confirm the efficacy data and to collect additional data on the complex interactions of organisms. The APHIS position indicated that tests can only be validated in the environment using standard agricultural practices and that such limited field testing is essential to develop a potential agricultural product. The stance was based on results of their review of the proposed field test and a subsequent finding that there was no risk of dissemination of a plant pest from the conduct of the test as described by CGI. The EPA also reviewed CGI's proposal (submitted as an application for an experimental use permit), as well as the local IBC; both of these reviewing groups were in concurrence with the APHIS findings.

It has become obvious that with increased public awareness, along with the existence of new and sometimes changing regulatory policies, tests such as the one illustrated here have taken on greater significance than is probably deserved. As the technology develops and field evaluations proliferate, companies are becoming increasingly concerned over the costs of the regulatory and registration process. In the case of the CGI field test, the experiments were carried out with local community input, which was generally supportive. The data being assembled from the test thus far indicate that there has been no adverse impact of the recombinant bacterium on the test site environment.

Overall, the regulatory agencies have tended to emphasize risk reduction in their reviews of cases involving the release of genetically engineered organisms in the environment with limited attention being given to the potential benefits, although the benefit of field testing to the development of information needed for subsequent regulatory decision making on the materials is recognized. Conditions for field testing thus far seem to stress the importance of designing field-release tests to control and monitor the dissemination of genetically engineered organisms from the site of release. The regulatory agencies can apply special conditions when needed to limit the risks associated with potential movement of the test organism in a field-release experiment, and they generally require plans for mitigating any unexpected damage that might occur. They also have the authority to limit or terminate an experiment when deemed necessary. The uncertainties, of course, still remain about field releases of genetically engineered bacteria, but the experience with bacteria engineered to harbor insecticidal properties indicates that we can probably expect that planned introductions of these microbes can be carried out safely.

REFERENCES

Adang, M.J., Firoozabady, E., Klein, J., DeBoer, D., Sekar, V., Kemp, J.D., Murray, E., Rocheleau, T.A., Rashka, K., Staffeld, G., Stock, C., Sutton, D. and Merlo, D.J., (1987) Expression of a *Bacillus thuringiensis* insecticidal crystal protein gene in tobacco plants. In: Arntzen, C.J. and Ryan, C. (Eds.) *Molecular Strategies for Crop Protection*, Alan R. Liss, New York, pp. 345–53.

Andrews, R.E., Faust, R.M., Wakiko, H., Raymond, K.C. and Bulla, L.A., Jr (1987) The biotechnology of *Bacillus thuringiensis*. In: Stewart, G. and Russell, I. (Eds.) *Critical Reviews in Biotechnology*, Vol. 6, Issue 2, CRC Press, Florida, pp. 163–232.

Anonymous (1988a) Ag bio update. In Wysiewicz, T.G. (Ed.) *Bio Engineering News*, Vol. 9, Port Angeles, Washington, pp. 1–4.

Anonymous (1988b) Biotechnology: managing the risks of field testing genetically-engineered organisms. In: GAO/RCED-88-27 *Biotechnology: Risk Management*, US GAO, Washington, DC, p. 108.

Aronson, A.I., Beckman, W. and Dunn, P. (1986) *Bacillus thuringiensis* and related insect pathogens. *Microbiol. Rev.* **50**, 1–24.

BARC Biosafety Committee (1988) Field testing behavior and fate of *Clavibacter xyli* subsp. *cynodontis*. In: Permanent Records of the BARC Biosafety Committee Project RD 076, Beltsville Agriculture Research Center, Beltsville, Maryland.

Biosafety Committee Project RD 076, Beltsville Agriculture Research Center, Beltsville, Maryland.

Barton, K.A., Whiteley, H.R. and Yang, N.S. (1987) *Bacillus thuringiensis* delta-endotoxin expressed in transgenic *Nicotiana tabacum* provides resistance to lepidopteran insects. *Plant Physiol.,* **885**, 1103–9.

Battisti, L., Green, B.D. and Thorne, C.B. (1985) Mating system for transfer of plasmids among *Bacillus anthracis, Bacillus cereus* and *Bacillus thuringiensis. J. Bacteriol.* **162**, 543–50.

Baumann, P., Baumann, L., Bowditch, R.D. and Broadwell, A.H. (1987) Cloning of the gene for the larvicidal toxin of *Bacillus sphaericus* 2362: Evidence for a family of related sequences. *J. Bacteriol.* **169**, 4061–7.

Calvin, D.D., Welch, S.M. and Poston, F.L. (1988) Evaluation of a management model for second-generation European corn borer (Lepidoptera:Pyralidae) for use in Kansas. *J. Econ. Entomol.* **81**, 335–43.

Carlton, B.C. and Gonzalez, J.M. Jr (1985) The genetics and molecular biology of *Bacillus thuringiensis*. In: Dubrau, D. (Ed.) *The Molecular Biology of the Bacilli*, Vol. 2, Academic Press, New York, pp. 211–49.

Crop Genetics International (1987) EPA application for an experimental use permit to ship and use a pesticide for experimental purposes only (Transmittal document). Crop Genetics International Corporation, Hanover, Maryland.

Dean, D.H. (1984) Biochemical genetics of the bacterial insect-control agent *Bacillus thuringiensis*: basic principles and prospects for genetic engineering. *Biotechnol. Gen. Engng Rev.* **2**, 341–63.

de Marsac, N.T., de la Torre, F. and Szulmajster, J. (1987) Expression of the larvicidal gene of *Bacillus sphaericus* 1593M in the cyanobacterium *Anacystis nidulans* R2. *Mol. Gen. Genet.* **209**, 396–8.

Ellar, P.J., Thomas, W.E., Knowles, B.H., Ward, S., Todd, J., Drobniewski, F., Lewis, J., Sawyer, T., Last, D. and Nichols, C. (1985) Biochemistry, genetics, and mode of action of *Bacillus thuringiensis* delta-endotoxins. In: Hoch, J.A. and Setlow, P., (Eds.) *Molecular Biology of Microbial Differentiation*, American Society of Microbiology, Washington, DC, pp. 230–40.

Fischkoff, D.A., Bowdish, K.S., Perlak, F.J., Marrone, P.G., McCormick, S.M., Niedermeyer, J.G., Dean, D.A., Kusano-Kretzmer, K., Mayer, E.J., Rochester, E.E., Rogers, S.G. and Fraley, R.T. (1987) Insect tolerant transgenic tomato plants. *Bio/Technology* **5**, 807–13.

Freedman, A.M. (1985) Rohm and Haas puts bug-battling gene into tobacco plants. *Wall Street J.*, 24 January 1985, p. 18.

Ganesan, S., Kamdar, H., Jayaraman, K. and Szulmajster, J. (1983) Cloning and expression in *Escherichia coli* of a DNA fragment from *Bacillus sphaericus* coding for biocidal activity against mosquito larvae. *Mol. Gen. Genet.* **189**, 181–3.

Gonzalez, J.M., Brown, B.J. and Carlton, B.C. (1982) Transfer of *Bacillus thuringiensis* plasmids coding for delta-endotoxin among strains of *B. thuringiensis* and *B. cereus*, Proc. Natl Acad. Sci. USA. **79**, 6951–5.

Graf, J.S. (1985) Mycogen gets EPA go ahead to test recombinant pesticide. *Gen. Engng News* **5**, 1.

Herrnstadt, C., Gilroy, T.E., Sobieski, D.A., Bennett, B.D. and Gaertner, F.H. (1987) Nucleotide sequence and deduced amino acid sequence of a coleopteran-active delta-endotoxin gene from *Bacillus thuringiensis* subsp. *san diego*. *Gene* **57**, 37–46.

Hindley, J. and Berry, C. (1987) Identification, cloning and sequence analysis of the *Bacillus sphaericus* 1593 41.9 kD larvicidal toxin gene. *Mol. Microbiol.* **1**, 187–94.

Höfte, H., Seurinck, J., Van Houtnen, A. and Vaeck, M. (1987) Nucleotide sequence of a gene coding an insecticidal protein of *Bacillus thuringiensis* var. *tenebrionis* toxic against Coleoptera. *Nucleic Acids Res.* **15**, 7183.

Klier, A., Bourgouin, C. and Rapoport, G. (1983) Mating between *Bacillus subtilis* and *Bacillus thuringiensis* and transfer of cloned crystal genes. *Mol. Gen. Genet.* **191**, 257–62.

Kronstad, J.W. and Whiteley, H.R. (1986) Three classes of homologous *Bacillus thuringiensis* crystal protein genes. *Gene* **43**, 29–40.

Obukowicz, M.G., Perlak, F.J., Kusano-Kretzmer, K., Mayer, E.J., Bolten, S.L. and Watrud, L.S. (1986a) Tn5-mediated integration of the delta-endotoxin gene from *Bacillus thuringiensis* into the chromosome of root-colonizing pseudomonads. *J. Bacteriol.* **168**, 982–9.

Obukowicz, M.G., Perlak, F.J., Kusano-Kretzmer, K., Mayer, E.J. and Watrud, L.S. (1986b) Integration of the delta-endotoxin gene of *Bacillus thuringiensis* into the chromosome of root colonizing strains of pseudomonads using Tn5. *Gene* **45**, 327–31.

Obukowicz, M.G., Perlak, F.J., Bolton, S.L., Kusano-Kretzmer, K., Mayer, E.J. and Watrud, L.S. (1987) IS50L as a non-self transposable vector used to integrate the *Bacillus thuringiensis* delta-endotoxin gene into the chromosome of root colonizing pseudonomads. *Gene* **51**, 291–6.

Oeda, K., Oshie, K., Shimizu, M., Nakamura, K., Yamomoto, H., Nakayama, I. and Ohkawa, H. (1987) Nucleotide sequence of the insecticidal protein gene of *Bacillus thuringiensis* strain *aizawai* IPL7 and its high-level expression in *Escherchia coli*. *Gene* **53**, 113–19.

Prefontaine, G., Fast, P., Lau, P.C.K., Hefford, M.A., Hanna, Z. and Brousseau, R. (1987) Use of oligonucleotide probes to study the relatedness of delta-endotoxin genes among *Bacillus thuringiensis* subspecies and strains. *Appl. Environ. Microbiol.* **53**, 2808–14.

Sekar, V., Thompson, D.V., Maroney, M.J., Bookland, R.G. and Adang, M.J. (1987) Molecular cloning and characterization of the insecticidal crystal protein gene of *Bacillus thuringiensis* var. *tenebrionis*. *Proc. Natl Acad. Sci. USA* **84**, 7030–40.

Vidaver, A.K. and Starr, M.P. (1981) Phytopathogenic coryneform and related bacteria. In: Star, M.P., Stolp, H., Truper, H., Balows, A. and Schlegel, H.G. (Eds.) *The Prokaryotes: A Handbook on Habitats, Isolation, and Identification of Bacteria,* Vol. II, Springer-Verlag, Berlin, pp. 1879–87.

Watrud, L.S., Perlak, F.J., Tran, M.-T., Kusano, K., Mayer, E.J., Miller-Wideman, M.A., Obukowicz, M.G., Nelson, D.R., Kreitinger, J.P. and Kaufman, R.J. (1985) Cloning of the *Bacillus thuringiensis* subsp. *kurstaki* delta-endotoxin gene into *Pseudomonas fluorescens:* molecular biology and ecology of an engineered microbial pesticide. In: Halvorson, H.O., Pramer, D. and Rogul, M. (Eds.) *Engineered Organisms in the Environment*, American Society for Microbiology, Washington, DC, pp. 40–6.

Whiteley, H.R. and Schnepf, H.E. (1986) The molecular biology of parasporal crystal body formation in *Bacillus thuringiensis*. *Ann. Rev. Microbiol.* **40**, 549–76.

Whiteley, H.R., Schnepf, H.E., Kronstad, J.W. and Wong, H.C. (1984) Structural and regulatory analysis of a cloned *Bacillus thuringiensis* crystal protein gene. In:

Ganesan, S. and Hoch, J.A. (Eds.) *Genetics and Biotechnology of the Bacilli*, Academic Press, New York, pp. 375–86.

Whiteley, H.R., Kronstad, J.W. and Schnepf, H.E. (1985) Structure and expression of cloned *Bacillus thuringiensis* toxin genes. In: Hoch, J.A. and Setlow, P. (Eds.) *Molecular Biology of Microbial Differentiation*, American Society for Microbiology, Washington, DC, pp. 225–9.

Whiteley, H.R., Schnepf, H.E., Tomczak, K. and Lara, J.C. (1987) Structure and regulation of the crystal protein gene of *Bacillus thuringiensis*. In: Maramorosch, K. (Ed.) *Biotechnology in Invertebrate Pathology and Cell Culture*, Academic Press, New York, pp. 13–27.

Index

Introduction of Genetically Modified Organisms into the Environment. Edited by Harold A. Mooney and Giorgio Bernardi

2002237586